FREEDOM AND RESTRICTION IN SCIENCE
AND ITS ASPECTS IN SOCIETY

CONGRESS PROMOTED BY THE
NETHERLANDS UNIVERSITY
FOR THE DISCUSSION OF THE SUBJECT

FREEDOM AND RESTRICTION
IN SCIENCE
AND ITS ASPECTS IN SOCIETY

THE HAGUE, 17 AND 18 SEPTEMBER 1954

THE HAGUE / MARTINUS NIJHOFF / 1955

This congress was organized by the Board of "Rectores Magnifici" (Vice Chancellors) of the Universities in the Netherlands.

Notable co-operation was received from the Delft and National Committee of "World University Service".

ISBN 978-94-011-8396-3   ISBN 978-94-011-9099-2 (eBook)
DOI 10.1007/978-94-011-9099-2

Speakers

DR. J. M. BURGERS, professor of aero- and hydrodynamics at the Technological University of Delft.

DR J. A. J. PETERS, C. ss. R., professor of metaphysics at the Roman Catholic University of Nijmegen.

DR R. KRONIG, professor in theoretical physics at the Technological University of Delft.

Pre-advisers

G. E. LANGEMEIJER, extraordinary professor [1]) in the introduction to the science of laws and philosophy of law at the State University of Leyden.

DR W. F. WERTHEIM, professor of the history and sociology of Indonesia at the Municipal University of Amsterdam.

DR H. W. JULIUS, professor of hygienics at the State University of Utrecht.

DR H. J. POS, professor of theoretical philosophy and the history of philosophy at the Municipal University of Amsterdam.

DR A. G. M. VAN MELSEN, professor in the introduction to philosophy, logic, logistics and natural philosophy at the Roman Catholic University of Nijmegen

Debaters

DR I. SAMKALDEN, professor in legal and political science at the Agricultural University of Wageningen.

DR P. J. BOUMAN, professor of sociology at the State University of Groningen.

DR M. G. PLATTEL O.P., professor in philosophical normative sociology at the Roman Catholic Economic University of Tilburg.

G. GONGGRIJP, professor in the oriental economics and the economic history of Indonesia, at the Economic University of Rotterdam; extraordinary professor [1]) at the Municipal University of Amsterdam.

DR M. C. COLENBRANDER, professor in ophthalmology at the State University of Leyden.

B. H. KAZEMIER, extraordinary professor [1]) in methodology and philosophy at the Economic University of Rotterdam.

DR D. WIERSMA, extraordinary professor [1]) of forensic psychiatry at the State University of Leyden.

DR C. J. DIPPEL, chemist in the employ of the "N.V. Philips' Gloeilampenfabriek" at Eindhoven.

DR K. KUIJPERS, professor in theoretical philosophy and the history of philosophy after the Middle Ages, at the State University of Utrecht.

---

[1]) Extraordinary professoriates exist in subjects which do not take a sufficiently important place in the academic teaching system to warrant the appointment of an "ordinary" (or full) professor.

CONTENTS

CONTENTS

INAUGURAL ADDRESS

by H. WAGENVOORT

Your Excellency,
Ladies and Gentlemen,

At the commencement of our activities it falls to me to bid you welcome on behalf of the Dutch Universities and on behalf of the Organizing Committee.

We are greatly rejoiced that our appeal to consider questions of fundamental importance for the development of Dutch science [1]) met with such wide response.

I particularly wish to address a few words of welcome to:

His Excellency the Minister of Statutory Trade Organization;
the Members of the Council of State and the High Court of the Netherlands;
the Boards of Governors of the Universities;
the representatives of the national and regional societies and organisations in the scientific field, amongst whom I may be permitted to mention in particular:

the Royal Dutch Academy of Sciences,
the Royal Dutch Institute of Engineers,
the Royal Dutch Society for the Furtherance of Medicine,
the Organization for the Research of Applied Physical Science,
the Organization for Pure Scientific Research.

His Excellency the Ambassador of the United States to his and our regret was prevented by official duties from accepting our invitation to attend this congress.

His Excellency the Minister of Education and the Governor of the Province of South Holland I rejoice to say, will be present this afternoon.

We, as well as he himself, regret that the Burgomaster of The Hague had to excuse himself. We appreciate it the more that the municipal authorities have been so kind as to offer us a reception in the old townhall this afternoon, at which the deputy burgomaster has expressed himself willing to do the honours. We wish to offer him our well-meant thanks.

---

[1]) "science" will be used to mean systematic and formulated knowledge.

To you, gentlemen of the board of vice-chancellors, who are really the chief personages to-day, I come last. For you after all are our principals and you speak through us.

I should like to express the hope that our work will not fall too far short of your expectations, and that at the close you may be able to look back on a successful enterprise.

The occasion giving rise to this congress will by now be generally known. The American University of Columbia in celebrating its bi-centennial this year not only put the theme "Man's Right to Knowledge and the free Use thereof" in the limelight of its own people, but it also induced the whole of the civilized world to consider that theme and the questions it raised. For the Netherlands this request reached the Senate of Leyden University, and its vice-chancellor, Prof. Duyvendak, was warmly in favour of the idea that in our country this and similar questions should be discussed in a congress. He brought the matter before the board of vice-chancellors and each vice-chancellor brought it before his own senate. That the reactions were somewhat divergent was especially the result of the fear some felt that a congress held at an American instigation would bear too little of a real Dutch character, whilst besides this many were of opinion that the subject needed elaboration. The theme Columbia had chosen might easily lead one to think of knowledge only as the result of the pursuit of science, whilst the problems one is faced with are inherent in its pursuance from the very beginning.

The board of vice-chancellors unanimously decided in favour of organizing a congress, whilst taking the wishes that had been expressed into account. It appointed an Organizing Committee and it was only natural that Prof. Duyvendak was asked to be chairman. Alas – soon after he started on the work, a serious illness seized him, from which he was not to recover. Not only his branch of science – the Chinese language and literature – thus sustained a heavy loss, and not only Leyden University, but our whole country and people. For on the one hand he was the acknowledged authority in his field and a credit to Dutch science, and on the other he was a most engaging personality. Now that we have to miss him here to-day I should like to request you to rise from your seats for a moment in order to devote your thoughts to him in grateful remembrance.

I thank you.

If you have perused the pre-advices, which have appeared in print, it will already have become clear to you that the *cause* of this congress is not identical to the occasion of it. The cause is rather that practising scholars and scientists again and again are confronted with difficult questions connected with freedom and restriction, – questions that do not only lay claim to their theoretical interest but whose urgence they feel as if in their own person.

I need not elucidate this; to-day and to-morrow these questions will be sufficiently discussed. One of the chief aspects that presents itself in this connection, to my mind the principal aspect, is that of the *responsibility* of the scholar and the scientist. We cannot yet speak of the lectures, but we find it running like a golden thread through the pre-advices from Langemeyer's first page to Pos' and van Melsen's advices that are purposely devoted to this term. It is consequently no wonder that pressure has been brought to bear on our Committee to include the word "Responsibility" on the side of "Freedom" and "Restriction" in the theme of our congress. It would have been entirely justified. It did not happen, however, because it was practically too late, because by the addition the correctness of the theme, it is true, would have been served but not its elegance, because it had already sufficiently appeared that the tension between freedom and restriction usually discharges itself in a critical consideration of responsibility. Indeed, now that our subject has been spoken of it perhaps deserves mention that lengthy discussions have been held on the question "can one speak of responsibility in science?" This question has apparently also engaged the minds of our pre-advisers and Pos, for instance, writes: "Our subject therefore must be so understood that it is not 'the arts and sciences' which may be called free or restricted or responsible, but only its practitioners." That of course is entirely correct, – at least if one takes the word "science" in its primary, abstract sense. But it may have a concrete meaning, in the way Julius takes it for instance when he writes: "Will it be possible to succeed in getting science, that is to say all its students or at least all those of any importance among them, to assume a certain unity as to shape and dimensions on the feeling of responsibility?"

The management of our congress believed that they should not apply any rigorous standards in this matter, being convinced that such a divergent usage of the word would exclude any risk of misunderstanding through the context in which it occurred. And I purposely used the somewhat pretentious term "management" a moment ago to show through another example how easily abstract nouns in Dutch can be used as collective nouns.

The importance of our congress would have been sufficiently elucidated in the foregoing, I hope, if it were not that stress should again be laid on its importance also as an expression of *national* and *international cooperation.*

This is the first time that all the universities in the Netherlands have jointly organized a congress, and also as far as I can see, the first time that they are acting collectively in external matters.

And even though this is not the first symptom of their having found each other – the many years of cooperation in the board of vice-chancellors proves the contrary – this is yet an important step that is a happy augury for the future of our inter-university community, and may therefore be considered of national consequence.

This is one of the reasons why we are so greatly rejoiced that Her Majesty the Queen has consented to act as Patroness of this congress, and that His Royal Highness the Prince of the Netherlands has agreed to accept the honorary chairmanship. For this we owe them our sincerest thanks. That Her Majesty has moreover made known her desire to attend part of our discussions fills us with gladness and pride.

In the international field our work also promises to bear fruit. The appeal Columbia University has made has met with response in many countries, so that there too the same problems will be discussed. To what this will lead cannot yet be estimated and there is no point in speculating about it. But it is unthinkable that the consequences of our discussions should not result in international consultation, and it is our hope that our board of vice-chancellors will in that case extend its activities still further.

I hereby declare this congress to be open.

THE VALUE OF SCIENCE

by J. M. BURGERS

I have been asked to speak to you on the value of science [1]) as an introduction to the subject of this Congress. Our experience concerning the present development of society and concerning its influence on human life makes us somewhat uncertain about the continually increasing part played by science in this connection. It may be useful therefore to pay some attention to the situation in which we find ourselves. The huge complex of questions with which we are faced is too large, however, to be treated by a single person. I can do no more than make an attempt to throw some light on a few aspects of the value of science, and in doing so I will stress the human side of the problem.

Everyone of you will admit that we are confronted here with a very fundamental aspect of human life. Present society, with all its various communications and its manifold contacts between people, depends in so large a degree on the application of scientific results, that this application has become indispensable. It has developed into a part of our daily routine.

This situation is accepted as something that is self-evident, and on the whole we do not pay much attention to the immense amount of thought and labour which has been required to produce it. An amount of thought and labour, moreover, which must be extended continually in order to preserve our status and to adapt it to the requirements connected with the steady increase in the number of human beings. To provide mankind with food, dwellings and clothing, the requirements of traffic and of communication, the health services, and the effective control of the many kinds of materials which are continually needed, is all dependent on the results of scientific investigation and cannot be executed without their help.

In this gathering it will be unnecessary to give an extensive picture of this situation. And it will be evident to you all that the methods and procedures applied, automatically require further development, refinement, and extension.

---

[1]) Here mainly used in the sense of the physical and natural sciences.

However, not only has science influenced and transformed our life in what is called its "material" aspects, science has at the same time provided us with large treasures for our spiritual life. Man has arrived at a better view of himself and of his place in the Universe, to an extent and a depth which was inconceivable before. A consciousness of mankind as a whole has become possible. This has exerted great influence on our thoughts concerning the meaning of life. Later on in this talk I shall come back to the meaning of science in this respect.

Hence the fundamental question: "Is it right to make use of scientific results, and have they any value for our daily lives?" has been answered in the affirmative by the facts themselves. Nobody will be inclined to turn back on this road, and we may use the term "homo faber" – man who makes use of instruments – as a biological term. Knowledge and its application form part of our biological make-up.

Thus our problem is not to be sought in the fundamental situation, for that is a part of our daily lives. No, it has to do with the details of the picture. The primary question before us is therefore,

Is *every* application of scientific results of value to us, is it useful, is it justified?

Along with this we must consider to what degree the application of scientific results influences man's mental attitude, and may drive him to commit deeds which are dangerous to himself or to others.

Once again, these two questions entail much which outsteps treatment by a single person. I shall not be able to give a formal answer to either of them. But in rough outline they describe the main features of the uncertainty to which I referred in my opening words.

One can say that there has always been some awareness, in one form or other, of this uncertainty, of possible dangers. Along with the dictum "knowledge is power", we are acquainted with discussions on the question "Is man allowed to eat from the tree of knowledge"? Many forms of scientific knowledge have at some time been decried as works of the devil. Only rather late in history, perhaps from the eighteenth century onward, appraisals of science make their appearance which do not recognize any difficulty.

Perhaps I may be allowed on this occasion to quote a few sentences from the original programme of Columbia University of May 31, 1754, as it was published by *L. C. Dunn*, in an article on "The Bicentenary of Columbia University" (Nature vol. *173*, pp. 703–705, April 17, 1954):

" ... to instruct and perfect the Youth in the learned Languages, and in the Arts of reasoning exactly, of writing correctly, and speaking eloquently; and in the Arts of numbering and measuring; of Surveying and Navigation, of Geography and History, of Husbandry, Commerce, and Government, and in the Knowledge of all Nature in the Heavens above us, and in the Air, Water and Earth around us, and the various kinds of Meteors, Stones, Mines and Minerals, Plants, and Animals, and of every Thing useful for the Comfort, the Convenience and Elegance of Life, in the Chief Manufactures relating to any of these Things; *And, finally, to lead them from the Study of Nature to the Knowledge of themselves, and of the God of Nature, and their Duty to Him, themselves, and one another, and every Thing that can contribute to their true Happiness, both here and hereafter* ....".

Evidently the idea governing this programme is that the study of nature can make us better, since it will help us to understand more fully how we should act towards God and towards our fellow men.

When we ask ourselves whether the expectations concerning the favourable influence of science on human behaviour have been realised, it will be necessary to differentiate our views. There are cases where the answer may be affirmative (perhaps this was felt somewhat more so in the nineteenth century); but there are also cases where a pessimistic judgment is called for. The cause of the diversity is the fact that in the human mind every scientific result is connected with effects of the driving forces of our life, so that the ultimate result appears as a product of many factors.

One should avoid separating the concept of "science" from the men or women who practise it. There is a tendency to personify certain notions; we use terms as, for instance, "to serve science", "science teaches", "science requires". However, in using such terms we shun the domain where an analysis is needed of the aspects governing the problem.

Science is made by *us*, men and women; we communicate our

results to each other, and we are the ones who make use of these results. All that follows from scientific research, like all human activity, is our own responsibility.

The spiritual value of science is embodied in the joy and in the widening of our understanding which are experienced on becoming acquainted with new relations. But this joy is coupled with the consciousness of greater power. The active attitude is already illustrated by the word "to comprehend". Of course, the joy of pure understanding may be preponderant, and the desire to apply the acquired understanding may restrict itself to the wish to penetrate towards still greater insight. But it is equally well possible that consciousness of an increased power for action shall be the preponderating feeling.

I am convinced that in any active person who is properly alive both aspects are primarily connected. Their separation is rather a consequence of education and of the influence of man's surroundings. Knowledge indeed *is* power, and it is characteristic of our mental structure that we wish to make use of that power. Science and technics are two aspects springing from the same side of our biological make-up, and are inseparably connected with each other.

Whereas in primitive times scientific knowledge and scientific techniques were often clad in a veil of mystery and secrecy, and were communicated to small groups of initiates only, the modern development of science, from about the 16th century onward, has been coupled with a growing consciousness of the personal value and the freedom of every man, and with the recognition of the fundamental equality of all men in regard to rights and duties. The idea grew that scientific activity was a thing which concerned mankind rather than the individual only. Automatically this led to the conviction that the results and views arrived at should be made available to everybody. From then onward results have been published freely, so that everyone would be able to obtain information and might consider what he could do with these results.

This open communication of all results has become of fundamental importance in modern science. Scientific investigation means penetrating into Nature's secrets along paths which continually become more difficult. What guarantee has man that he is following "right" paths, that his investigations will lead towards a better understanding of truth about phenomena presenting ever stranger aspects? Is there no possibility of losing one's way? The conviction that it is given to a single person to recognize what is real truth, may be a guide in a number of cases; but it is impossible to rely upon it in an absolute way. Too many notions which once had seemed to be beyond all doubt have later appeared to be incomplete or false. Hence our sole hope must be found in co-operation and free criticism, in the expectation that unrestricted comparison and repeated investigation will in the long run enable us to eliminate whatever is insufficient or untrue, and that the results of joint deliberation will lead to an acceptable approach to the secrets of Nature.

This is the conviction in which we are still living. The only guarantee for keeping science "on the right path" is to be seen in the free development of individual minds and in team-work springing from free co-operation. Every infringement of this freedom, every restriction in the possibility of communicating results, every enforcement of a way of thinking or of treating problems, threatens the purity of our search for truth.

In principle this requires that every man, in every part of the world, must have the possibility to acquaint himself with all current opinions and with all available results, so that he may think them over and test them himself. It will be evident that this necessitates special requirements concerning the way in which scientific results are interpreted (and, I should like to add, concerning the material aspects of publication). The main point, however, is that this unrestricted communication embodies the ideal democratic conception in science, which itself is a fruit of the ideals of western society. The search for insight and for truth finding expression in science represents a fundamental activity of human life.

I have made use of the terms "in principle" and "ideals". Indeed, the question may be brought forward whether the notion of "every man", taken in an absolute sense, does not lead us too far in this connection? There is a general idea that children need not or even should not know everything; along with "children" one is wont to speak of "immature" people. There are men and women who are so restricted or so muddled in their thinking, that exchange of thought with them on scientific problems cannot lead to results. Surely there can be no obligation to explain everything to them.

There are also persons whose mental powers are disturbed, and with an obvious extension one can pass on to people in whose mental make-up or in whose character there are serious faults. The question then arises, what will they do with the knowledge that comes to them?

There is real danger when big forces are at the disposal of immature people. Fire-arms and motor vehicles give sufficient examples, but also data expressed in words can be used to inflict damage on others.

Perhaps we may amend the rule concerning the availability of scientific results for everyone in the following way. Everyone should be free to become acquainted with all results, provided that they have the necessary basic knowledge and provided they are trustworthy. The term "basic knowledge" is not so far-reaching in this respect; but "trustworthiness" embodies a concept which is liable to great differences of opinion. One might attempt to define the term as "whoever is guided by the same high ideals of truth, humanity and co-operation that we all acknowledge", but it is evident that every word of this phrase can be interpreted in many ways.

Social difficulties, in particular the events of the last years with their fierce opposition between two contending parties in the world, have forced this problem acutely on our attention. The notion of "trustworthiness" in several cases has been interpreted approximately as "whoever is impervious to any other thought than such as is considered imperative by certain groups". The dangers of such an interpretation will be evident. In a fatal manner it overlooks the fact that loyalty is something different from conformity. Each form of society must be built upon the co-operation of minds of different types, and each form must be open to development, under penalty of deteriorating. It is wrong to fix the possibilities for development of a person's thoughts beforehand.

However, it is neither possible, nor is it in the plan of this meeting, to enter fully upon questions of this type. We recognize that no final solution is possible. Under the present circumstances we cannot give the same rights to *all* people without distinction, but nobody has the necessary wisdom to mark the dividing line valid for all cases. There is no other way than to struggle never-endingly. Our hope must be found in the observation of a sense of equity and of fairness in the contending parties, so that dividing lines can remain mobile, and can continually be adapted to new conditions.

Even if we leave this difficult subject aside, there is still a wide field of problems before us that concerns the influence of science on society. We shall assume, therefore, that in our present society the great majority of people are considered as "mature", and that, with the exception of some rather well-defined domains, scientific results can be freely published, and that the same applies to many data concerning their application.

In our western society, based on free enterprise, this situation entails that arbitrary persons are allowed to make use of the results of science according to their own judgment, subjected only to certain general rules and laws as prescribed by society. Until now we have always considered free enterprise as an aspect of democracy and as a condition for unhampered development.

In the times when these ideas arose, the activity of one man reached only a limited group of other people, so that problems of admissibility and of use could be judged on relatively simple grounds. However, with the growth of science the bearing of its applications has increased enormously. Bitter experience has taught us that applications may lead to results completely different from what had been meant originally.

Society has proved to be unstable against the powers which science and technics have placed in man's hands. Transformations have appeared which have brought disaster to numerous groups. Much has been washed away of what we once considered to be of value, while no new values have taken the place of the old ones. With nearly every application effects have made themselves felt which had not been foreseen. In other words our faculties of foresight and of organisation have not corresponded to the powers which we were given to handle.

To mention a few examples: At the beginning of the industrial revolution the rules regarding property and right of control were such that there was no responsibility concerning many results of one's activities, in particular this was not so in respect to indirect consequences. Furthermore, he who once had started to gain economic power could continue almost indefinitely, even when his power inflicted curtailment of freedom on others. There were no rules which revealed the use of power, of energy, and of natural resources as a common heritage of mankind, which should be controlled in view of a more distant future than some years or some decades only. Who thought about rules concerning the deposit of refuse? Who paid any attention, in the beginning of the industrial revolution, to the principle that work is essential to man's life and that work in industry should be organised in such a way as to contribute to the development of personality? What has been left to us of peace and quiet?

All social and political rules, laws and modes of thought were based on the idea of a relatively stable, slowly developing society, in which the consciousness of the common fate of all mankind played an insignificant part. The result at which we have arrived is completely different: we live in a rapidly changing, often anarchistic, and in many respects destructive world.

It is sometimes said that technical development has generated forces which can no longer be controlled. These "forces", however, are not the powers in fuel or minerals, nor those in electricity or nuclear energy. The "forces engendered" are the desires in the hearts of men, driven as man is by expectations, by hope, by fear, and strongly influenced as he is by problems of security seen only from a very narrow point of view, while being swayed by emotions, which he then proceeds to formulate as his profoundest convictions.

These forces have been focussed into sharper consciousness and greater potency by the growth of technical possibilities; they have been strengthened by our communication system, which makes possible a close contact between all peoples and enables man to produce mass effects of world-wide extent. Such mass effects can be obtained in particular when the outlook is narrowed and when all instinctive energy is driven into a single channel.

In this connection it is important to observe that the human mind is often inclined to restrict its attention to certain lines of development. This feature is connected with the basic character of our intellectual faculties, which find their greatest power in *analysing*, that is to say, in splitting up a situation into separate aspects, and in considering every aspect on its own merits. There is a limitation in our aptitude to comprehend many aspects of what happens around us simultaneously. Each one of our conceptions realises a finite perspective of life only and leaves other aspects unnoticed.

In scientific investigation this method of attack has proved to be extremely fruitful. However, in life neglect of the connections with the surrounding background may have fatal consequences. One may say, moreover, that when the urge towards activity is stronger, the one-sidedness of our mind becomes more marked. It is true that we attempt to make "syntheses", by combining a number of aspects, each of which is obtained by abstraction and analysis, but it has become evident that in the fields of technics and of social organisation our present syntheses often do not guarantee a harmonic unfolding of the deeper values of life. All that we undertake remains fragmentary. There is, moreover, a dangerous urge in our souls, apparently arising out of our biological structure and connected with our tendency towards concretisation, namely the urge to fight to the bitter end, to remove obstacles and conflicts at any cost. This urge may ultimately lead to the desire to destroy.

In particular the authority of quantitative data may obscure the sense for harmonic development. Although the science of mathematics is a wonder of beauty and of comprehensiveness, nevertheless quantitative ratios obtained in the investigation of the phenomena of life can illustrate no more than certain disjointed aspects.

We are faced here with a problem where our methods of scientific research, through their stress on quantitative relations and their elimination of judgments concerning values, cannot entirely be acquitted of one-sidedness. It is necessary to keep this in mind when considering the value of science for our everyday lives. Do not think that I should like to reject these methods in science, I know too well what they mean to us and I know the beauty that can be revealed by numbers. But it cannot be denied that their continuous application has given rise to habits of thought inducing us to give a preponderant position in our lives to counting and measuring. This is particularly the case when social phenomena are studied from an economic point of view. A similar feature is present when we believe that democracy can be fully realised by counting votes, without considering the various ways in which convictions are formed, nor giving attention to the value of individual expressions of opinion.

There is a danger of a lack of creativity when it comes to coping with the enormous increase in the quantity of our production. There is a danger of insufficient inventiveness in human relations. Science itself undoubtedly represents a domain where creative thought has worked to a very wide extent – but it is a domain which embraces only part of the activities of life. We cannot deny that the success of scientific quantitative thought and the elimination of judgments concerning value, which is a "conditio sine qua non" for science, has been a contributing factor in the development of a trend towards giving preponderance to counting or measuring instead of judging, when dealing with social problems.

However, I must be careful, it is possible that I may be producing a wrong impression. There are more sides to the problem. Scientific investigation in many cases has shown that old rules for judging values were based on fallacious ideas, lacking any real foundation. The elimination of such obsolete rules meant progress; it took away the obstacles that stood in the way of the formation of a free opinion. But science itself does not furnish any opinion on value; it does no more than open the way, and it leaves the conception of a new valuation to other faculties of our mind. The error which I tried to bring out was that we could not harness these other faculties into proper activity. Moreover, in this process we have put ourselves too much in opposition to nature. All this has caused a lack of unity in our lives and a certain anarchy may be observed in much that has been brought about by modern technical development.

With this group of problems I touch upon fields of philosophic and religious thought, which are not my province and which again outstep the limits of this Congress. It will be unnecessary to remind you that there are other sources of mental life in us than the desire for knowledge. We may ask whether the negative influence of science to which I have alluded is the only one; and whether or not, in view of those other sources of mental life, science has also contributed to the *positive* side and has helped us to conceive new and better judgments about value. Perhaps it depends on social relations as well as on personal inclination, which of the two sides of the picture receives most of the light; in other words, which side finds the deepest resonance in us.

I will mention a few of these positive influences, where science has enriched our world picture and helps us to arrive at a more comprehensive valuation.

Science has revealed a diversity of relations between features and events, in Nature and within ourselves, which have far surpassed our naive ideas. The notions of space, time, presence, being and existence have been changed and transposed; and they present different aspects when we are dealing with different phenomena. All these relations must have their ultimate meaning in God. Hence we may say that our idea of God has become enriched.

It has been found at the same time that these new relations are not mere abstract speculations: many of them influence concrete practical occasions. Hence, when reasoning according to what is called "common sense", it is necessary to pay attention to these new types of relation, since they appear more often and exert a greater influence than had been suspected until now.

In particular, new light has been shed on the opposition between "object" and "subject". In the investigation of "lifeless phenomena" along the lines of physics as well as in the investigation of the activities of life (including those within ourselves) it has become increasingly clear that there is no absolute separation between subject and object, and that the observer and the observed influence each other reciprocally. Not only that the observer learns from what he has observed, but that which is observed has experienced the influence of the observation. Every observer, indeed every one of us, is continually "on the inside". Hence we no longer stand over and against nature. We have learned to understand our position better and we are conscious of partaking in the whole.

Another instance: To become acquainted with the wonders exhibited by microscopic organisms, to gain insight into physiology and psychology, widens our own consciousness of life. Such insight can give us joy and happiness through the beauty which is revealed to us, and at the same time it can lead us to greater patience, to better understanding and to milder judgment in dealing with our fellow-men. Moreover we have learned about human desires and trends, and we have been able to sound tendencies towards self-destruction. We even hope to find forms of protection against such tendencies.

We have gained insight into the possibilities of the transformation of matter and energy. The concept of matter has lost its character as an absolute; it no longer appears as something of a lower order, something inimical to "mind". The concept of matter has ceased to represent something static: "matter" must be conceived as the historical route of a chain of processes which repeat a certain pattern very many times. Perhaps there is, in principle, no fundamental difference between an atom or molecule many times repeating a given set of electronic movements, and, say, a state or a human community, which in consequence of its system of laws and customs exhibits a permanent national character during many generations. Once this insight has been gained, matter loses much of its refractoriness; and there is no longer a desperate antithesis between matter and mind. Our spiritual aspirations can express themselves more freely than ever before.

Furthermore we have become acquainted with the finite dimensions of the Earth and the finite magnitude of the natural resources available to us. The essential unity of the human race begins to obtain definite features in our consciousness and we recognize that we are taken up in an adventure of life, which concerns all of us together. We are beginning to attain a better view of our responsibility with regard to posterity, to whom we must transmit good mental and bodily faculties and a good environment.

Along with the finite extent of space on Earth we have become conscious of a time scale which has to be counted in millions of years. It is on this scale that we have to project our future development. The deeper historical insight which results from this time scale, together with its picture of evolution, may help us to recognize the relativity and the temporariness of many of our present acquirements.

At the same time the problem of the number of men, who can

live on Earth and strive for truth in harmony with each other, begins to present an acute form.

Everywhere the interdependence, of men and their fellows, of men and other living beings, animals and plants, and of men and non-living nature around us, is forced upon our consciousness. "All partakes in all" – these words of the 19th century Dutch writer Multatuli may be repeated here; indeed, it may be that all nature together forms only a single conscious Being, of which we are no more than the highest organs. If it is possible to improve these organs, if we can realise a new generation which will be able to understand God more fully, it will be our task to do so, and knowledge, science, in the widest and in the most human sense of the word, will be one of the great helps.

Such considerations bring us to a new consciousness of our responsibility, responsibility with regard to posterity, responsibility with regard to the whole of our surroundings, responsibility with regard to God, viewed in the light of a space which encompasses the whole Earth (may be within some time also other celestial bodies), and which extends throughout an indefinite length of time. This consciousness of responsibility must find its expression in a new form of ethics, combining the fundamental values of humanity and honesty with the far-reaching relations that have been uncovered by our investigations and which open ever-widening perspectives.

What the new ethical values will be cannot be foretold and a theoretical construction has no sense. It is even possible that we are already engaged in re-modeling our biological make-up, and if not now, this may be the case within a short time; it will then be our task to make a new synthesis. A new applied ethics can be made only in actual life. However, when we take as starting point our responsibility with respect to the coming generations, some practical conclusions can be drawn at once. We must leave to these generations surroundings which are not restricted as regards the possibilities that are left open; an Earth which is not bereaved of her beauty, nor empoisoned by our refuse, nor by the remains of an excessively destructive apparatus of war; and the hereditary material which we leave to the future must not be reduced to shallowness, nor must it be crippled by fears and repressions. We must give to our children and our children's children the means for realising greater freedom of self-expression liberated from our tendencies to self-destruction, and the possibility for enjoying the beauty which nature and artistic sense can give us.

Increase of responsibility will entail a restriction of what once appeared to be freedom. New forms of freedom, however, will take the place of the old ones. Education is necessary to ensure the proper adaptation to this new responsibility and to prevent it from deteriorating into a system of compulsion. For this purpose we do not primarily need education in scientific thought; what is needed in the first place is good human relations, helpfulness, care and attention, and recognition of each others personality, supplemented by good general intelligence. Notwithstanding the impressive development of knowledge and technics in the medical field, to take one example, there will always be a need of nurses who are deeply conscious that their calling requires devotion to the needs of other people. Something analogous might be said of many fields, and it is certain that science will only find its full value in human society against a background of human concern and devotion, added to a consciousness of responsibility with regard to other people, nature, children and their children.

Finally: one of the wonders of science is its continuous "auto-correction", its growth through the never ending re-working of its conceptions and images. It is necessary that more attention be given to this feature, in society as well as in education, so that also there change will be recognised as a necessary accompaniment of growth. A stable principle of co-operation can only be reached when the temporary nature of every form of government is recognized, so that one shall be able to alter that form each time the rapidly changing aspects of technical possibilities face us with new problems.

The same as in science, this necessitates *openness* in our relations. The "auto-correction" of science – it has been mentioned before – is possible only when every tendency towards secrecy is eliminated. This should serve as an example for social relations, in which we must strive towards the same candour as is to be found at the base of all scientific work. In science and in society it is impossible for a single person to discover what will be good and useful in the long run; and open criticism and co-operation is needed in order to eliminate what is wrong before it leads to disaster.

In view of this it will be of the greatest importance to make investigations concerning the ways along which results of scientific investigations are introduced into society. Between the work of the scientist in the laboratory or at the writing table, and the man who makes use of new products, new apparatus, new means of travel or communication, there extends a long chain of decisions which for a large part are governed by commercial considerations, taken by persons and by enterprises having no responsibility in this particular sense. The far-reaching influences which such decisions may have on the structure of our society, do not count in these decisions. Whoever it was who introduced the light motor vehicles on the market, was not required to give a public justification of this deed, and nevertheless, an endless series of traffic problems has been raised by it! The enterprises which put the radio at the disposal of such an enormous public, have put mankind before educational problems of an extent never known before. In the long run mankind cannot allow important decisions affecting our surroundings and our mode of life to be taken in a manner which does not permit a timely appraisal of such consequences. It must become possible to influence or to prevent a decision, if the necessary adaptation cannot be assured in time.

It will not be a simple matter to find ways and means for this; the relations are quite different from those applying to science, and man is not free from a desire to hide things from others. But we shall not escape from the necessity of devising certain forms of public responsibility with respect to much which at the present moment is still considered to be the exclusive right of private enterprise.

Only in a society which is prepared to investigate its own structure and which in all its relations and activities strives after candour, can science come to its full value – this is a complement to what I have said about the necessary background of humanity and care.

I should now like to be permitted to quote a few sentences from a speech made by General *J. C. Smuts* before the Centenary Meeting of the British Association for the Advancement of

Science, in 1931 in London, on the "Scientific World Picture of To-day". The following paragraphs have made a deep impression on many people (Report of the British Association for the Advancement of Science, 1931, The Presidential Address, p. 13):

"Among the human values thus created science ranks with art and religion. In its selfless pursuit of truth, in its vision of order and beauty, it partakes of the quality of both. More and more it is beginning to make a profound aesthetic and religious appeal to thinking people. Indeed, it may fairly be said that science is perhaps the clearest revelation of God to our age. Science is at last coming into its own as one of the supreme goods of the human race.

While religion, art and science are still separate values, they may not always remain such. Indeed, one of the greatest tasks before the human race will be to link up science with ethical values, and thus to remove grave dangers threatening our future. A serious lag has already developed between our rapid scientific advance and our stationary ethical development, a lag which has already found expression in the greatest tragedy of history. Science must itself help to close this dangerous gap in our advance which threatens the disruption of our civilisation and the decay of our species. Its final and perhaps most difficult task may be found just here. Science may be destined to become the most effective drive towards ethical values, and in that way to render its most priceless human service. In saying this I am going beyond the scope of science as at present understood, but the conception of science itself is bound to be affected by its eventual integration with the other great values".

I do not know whether I have given you a satisfactory picture of the value of science. I am too much influenced by a consciousness of this value than that I feel able to give a definition of it. What I have tried is rather to give you an impression of the human relations which are its determining factors.

The original theme proposed by Columbia University was: "Man's right to knowledge and the free use thereof". I hope that you will be convinced that this right has been given to us at birth, but that all depends on what use we make of it. All factors giving meaning and direction to our lives play a part in this connection;

and everyone of us has a share in the responsibility for what we do with our lives and with our knowledge.

In finishing I should like to stress my conviction that all persons occupying leading positions in our society – whether political, economical, industrial, legal or educational, the greater part of whom are not scientific workers themselves – nevertheless have an important and never-ending task in the interpretation of our responsibility and in the development of new ethical values.

ADDRESS OF WELCOME TO
HER MAJESTY QUEEN JULIANA

by H. WAGENVOORT

Your Majesty,

At the re-opening of this meeting I have the pleasure of bidding you a sincere welcome on behalf of the Board of Vice-Chancellors of the Netherlands Universities and on behalf of the Committee whom they entrusted with the organisation of this congress. And I also wish to express our joy that you have not only declared yourself willing to act as Patroness of this congress, but that you have also been so kind as to give dephts and splendour to our gathering by your personal presence. We are very grateful to you for this, in the first place because we can salute in you the Queen of our country and of our hearts. But not for that only. Your Majesty, who yourself went through an university training and show an attachement to your Alma Mater that might act as an example to many, are driven of that we are sure – by a warm interest in the weal and woe of science and the problems with which it struggles. And in the third place, as I had occasion to remark at the opening this morning, the central problem of this congress "Freedom and Restriction in Science" brings along with it that again and again, and very particularly, stress is laid on the necessity for a sense of responsibility in the individual and in the community, in the student of science [1]) as well as in science as a whole. There is also a science of responsibility, and who is more familiar with it than you who perhaps do not give lectures in it but who daily give an exemplary practical demonstration?

We gratefully mention the vigorous support herein given you by H.R.H. the Prince of the Netherlands, who has been so kind as to honour us by acting as Honorary Chairman to this congress.

It is for these reasons, Your Majesty, that we salute you, who have been placed above us all, as the one from amongst our own number whom we should most prefer to see here to-day.

---

[1]) "student of science" will be used to mean practising scholar or scientist.

Finally I should like to welcome his Excellency the Minister of Education and the Provincial Governor of the province of South Holland. They may be assured that we greatly appreciate their presence.

THE LIMITS OF SCIENCE

by J. A. J. PETERS

Your Majesty, Ladies and Gentlemen:

Science can be understood both in the subjective and in the objective sense. When we take it in the subjective sense it is quite a special manner of *knowing*, when we take it in the objective sense it is quite a special whole of *things known,* precisely in so far as these are known by that scientific manner of knowing.

It is not our intention here to go into the fundamental relation between the subjective and the objective side of science, between the knowing and the known. To raise the problem of limits of science it suffices to consider the obvious *fact* that in science there *is* such a thing as a *tension* between the knowing subject and the object known. Neutralizing this tension by integration of the object into the immanence of the subject would no doubt be the realization of ideal science. This *ideal,* however, viz. the subject comprehending being-as-a-whole through a complex of critically verified judgements, will never be attained by man as long as his existence is historical and developing in time. Science, therefore, has beside its sense as an ideal, especially the sense of a *historical reality:* the whole of critically ascertained certitudes at any particular moment. We can compare this whole of certitudes with a shining nucleus, surrounded by a vague halo of theories and hypotheses trying to explain, to understand and to connect these certitudes.

Though, then, the ideal of science is not attainable at any moment, yet it stimulates the investigator in his search. It happens indeed, that at definite moments of history knowing man is less aware of the distance between science he has attained and the ideal of science. He then prematurely *thinks* he has arrived at a *definite* conception of being-as-a-whole or of a special realm of being. That is the exuberance of youth.

It also happens, on the other hand – and this belongs to the ripening process of science –, that science meets with unexpected obstacles on the way, or all of a sudden, as if intuitively, takes an unexpected turn. The wonder at the process of science then stimulates the investigator to *reflection.* He checks the impetus of his directedness towards the object for the very reason to *measure the distance* between the attained and the ideal, between the historical appearance and the essence. In doing so he becomes aware of the actual limits, finds a further horizon dawning, new realms still unexplored or new untrodden ways of explanation.

Such a critical reflection also exists when the specific sciences – already in a far-advanced stage of independence and of successful application to special fields of life and technique – want to strengthen *the bond with the whole* and either ask for their common root by examining their foundations or expect a synthesis of their results in a worldpicture by deepening and widening the leading theories.

Finally, such a reflection is also needed, when in connection with the increasingly responsible task, entrusted to the sciences in the technical and economic, the social and cultural life, insight is wanted into their essential *value* for the guidance of these realms of life.

When we are going to put the problem of the limits of science, we are not going to talk about the limits that – rightly or wrongly – are set to the *spreading* and *application* of scientific observations and insights (and consequently perhaps also indirectly set to the very process and direction of the research). Such limits may arise from demands – reasonably justified or not – of the economic, the social and the ethical life. This important aspect of the problem will be dealt with when the main theme of the congress comes up for discussion. We, however, are going to confine ourselves to the question whether science itself *in its own autonomous development* would, on reflection, come to the acknowledgement of its limitedness. The question which concerns us, therefore, is: whether there are any *intrinsic* limits of science. In other words, limits to be acknowledged not by a superscientific authority, but by science itself i.e. by seeking man in critical and freely accepted responsibility. The point for us is to get insight into what, *in principle*, he can or cannot do and to consider what *in fact* he did or failed to do when seeking after truth.

Whereas the previous speaker brought out the value of science, we shall– in accordance with the task we accepted – have to put forward its misery rather than its grandeur. But even in that misery we want to stress the grandeur of science; for it is the man of science himself who faces the limitedness of his science. For to be able "to acknowledge" one's own limits is altogether different from merely "having" these limits. This awareness of limitation, being already a certain transgression of limits, forshadows a certain unlimitedness. Only when directed towards a farther, clearing horizon, man is able to discover the limitedness of the attained or in this way attainable. So when searching man has become conscious of the fact that he is finite in his search, a vague and confuse *perspective of a more comprehensive science* is revealed.

But even within the very bounds of science its grandeur manifests itself. For as a more perfect form of knowledge science is also directed towards truth in a more perfect way. The possession of any particle of truth is of infinite moment. *Any knowing* is as such always a *good* by itself, which it is better to possess than not. Should someone say, that in certain circumstances it can be better not to know certain things than to know them, I would answer that this can never be so on account of knowing as such, it can only be on account of circumstances that lie outside knowing as such. The value of knowing is unassailable.

The first thing we need here and now is an accurate *definition* of the concept science. It is, however, by no means simple to define science as a particular and more perfect form against the background of knowledge in general, precisely because of the historical character of real science. For even in its self-knowledge science is growing and so still incomplete. This does not alter the fact that it keeps on trying to understand its essence, which is present as such in every alteration, in every variation.

Knowing in the usual sense is opposed to guess, opinion and belief and indicates possession of a truth with certainty. This wide sense of knowing, however, is unsatisfactory to explain what we understand by *science*. In other words science has a more specific sense than knowledge in general, nor is knowing-with-certainty a satisfactory definition of science. With the Greeks the word "epistèmè" was, as a rule, restricted to a special form of knowing-with-certainty. Think of Plato, who at least in the Politeia, book 7, places "epistèmè" above discursive thinking and relates both to the essence (ousia) in contrast to "opinion" which concerns the world of becoming. Think also of Aristotle who in Analytica Posteriora I,2 gives the well-known definition: "There is only question of science when we know the cause why something is as it is, but also that this cause is the cause of this something and that this something can but be as it is". However numerous the shades of the concept science may be in the process of development in Antiquity, a clear distinction was made between knowing in the vulgar sense as used in everyday life and that form of knowledge, more refined, of a higher order and pursued for its own sake, called science.

This distinction does not in the first place lie in the object as such, but *in the degree of critically verified certainty;* when Plato and Aristotle think that universality and necessity are characteristics of the *object* of science (and therefore do not acknowledge e.g. history to be a science), further development has led to an expanded concept of science in which universality means universal accessibility for all *subjects* and necessity means the undeniability of whatever is, be it contingent or free.

Though *pre-scientific knowing* is based on some unquestionable certainties, it is not possible to make on that level an exact distinction between this hidden germ of truth and the aggregate of more or less primitive opinions and prejudices, in which this germ is couched: antropomorphisms and perspective misrepresentations and misconceptions from the too-limited viewpoint of individuals or groups, aiming directly at utility for life. Clashes between cultural unities of a different mentality usually give rise to a *crisis*, in which a distinction is made between what inevitably belongs to the very essence of things, being open to observation and reason, and the mythical and practical views which vary in different milieus.

In this way methodical reflection aims at stripping vulgar knowledge of its deforming subjective factors and at reaching the higher level of objectivity and verifiability. The guiding principle of this reflection is no longer utility for life, nor emotional preference, but the possibility of man to be open to and turn freely towards *what is*, as it reveals itself to the unbiassed observer and thinker. Not unly purified experience, but also purified reason plays a part in science. For isolated knowledge is always imperfect knowledge, as every datum not only points to other data, but also implies presuppositions and consequences. If man wants to know what he knows, then he will also have to bring to light explicitly these real and logical connections – and with that an infinite field of research demanding arrangement of concepts is disclosed to him from any starting-point in experience. In this way science comes about and grows into a coherent whole of concepts, judgements and reasonings. This then is a new characteristic of science, beside *love for truth-for-its-own-sake*, beside *universality* and compelling *undeniability*, beside *critical method:* science urges towards forming a "system" under the aspect of both formal logic and material content.

Out of the long historical development of sciences towards self-realization most of our attention must be given to the fact, still so important, that beside philosophy as an attempt to embrace being-as-a-whole, a growing series of *specific sciences* has attained a growing degree of *independence*. The Ancients, too, sometimes assigned a private place to mathematics, astronomy and geography, to some other scientific technics and arts. But it is the great revolution of modern times, that the physical sciences have reached full development as to both their methods and their contents, by refinement of observation in the experiment and by using mathematics as an auxiliary science. And after the sciences of *nature* it seems that those of *culture* at last have found their way, at any rate as to their methods. For much greater are their difficulties to come to universality and systematic construction, because of both the "wealth" of the object and its singleness and originality, by which it is less universally accessible and verifiable.

When we put the question of the limits of science, we henceforth have in mind the position of the *modern, positive sciences*, which – in spite of all variety in their specialized and combined methods and contents – have their poles of attraction on the one hand in *physics*, which are directed towards the recurrent and the universal and in this way capable to give us some certainty as to the future, and on the other hand in *history*, which is directed towards the irrepeatable and the individual and makes us aware of the past. Their common origin, however, is an intensified urge to a certainty that binds everybody in unprejudiced objectivity.

Objectivity here even gets a very particular sense: the directedness of the subject towards an independent datum facing him which he wants to describe and understand exactly as it is in itself – independent of his observation. By this exclusion of all the factors and influences on the part of the subject these sciences try to effect that they are not only accessible to this or that mentality, but that they *are open*, at least in principle, *to all mentalities* – so irrespective of every particular mentality. The way, therefore, along which definite "wholes of known objects" have been reached, is part and parcel of science itself. Not only things arrived at, but especially starting-point and intermediate links must be communicated so that others might examine whether in fact no "subjective" elements crept in during this process of observation and thinking and whether the investigator was only led by the demands of the object. That is why through precise observation and close argument all sciences try to realize thus the ideal of science: contents of knowledge unified into a system, universally accessible and communicable, the truth of which can be checked against the undeniable facts.

But if science reaches full development only in self-criticism, then it has, trying to realize the adaptedness of its methods and contents, also put the question of possible limits. If science wonders why what is essentially accessible through science in fact escapes science in the present state of research, then it observes imperfections on the part of the knowing subject, which may be overcome by a purer attempt and by working along new lines; a new perspective opens on the level of empirical science itself. Science, however, may as well wonder whether objective knowledge, accessible in principle to positive science, is covered by the whole object of possible human directedness to truth. And then we should have to accept essential limits of science, if the very object of science proved to be pointing to what only might be attained by *other* manners of knowing, by knowledge that presented itself in other relations between subject and object, – in technique and art, morality, philosophy or religion.

In our observations here we shall have to change over from one problem to another, from more peripherical and factual to more central and essential limits of science.

That science must be unprejudiced means that it must debar deforming, subjective influences. But pre-scientific life is full of "idols", to quote the words of Francis Bacon. There are the common anthropomorfic conceptions, the idola tribus; there are the presumptions proper to everybody in particular, the idola specus; the idols that have crept into common parlance through intercourse, the idola fori; and idols brought about by tradition: the idola theatri. Our unwillingness to clear ourselves of all these false or critically unverified ideas commonly proceeds from *passion* – a passion different from the passion for truth. The history of the sciences is full of erroneous ways, owing to lack of self-control. Science willingly or unwillingly suffered subservience now to the preservation of the existing order, now to the justification of the revolution. Documents were made to disappear or were forged, statements and resolutions were misinterpreted, hypotheses too readily mistaken for evidence. The wish was father to observation and thought. At least in certain realms of science, the "quod volumus libenter credimus" can apparently only be escaped by radical self-criticism.

Besides hope *fear* played its part – fear of truth. It made people close their eyes, prevented them from seeing what could have been seen since long, acted as a check upon many a research. They forgot that truth will make us free and that there is no need for any view of life, justified in its own field, to entertain a fear of scientific observation and insight, provided the one as well as the other acknowledges its own nature and limits. However hard it may be in many cases to determine what the real meaning is of a conviction established within a philosophy of life, a religion, a Revealed Creed, and what the undeniable assertions and proofs of science really claim, it does not absolve us from our duty to give the one as well as the other its due – without fearing any particle of truth, no matter who discovered or revealed it.

Another passion that seems to be conducive at times, but in reality often works to the opposite effect, is the *recklessness* that urges towards originality – by this is not meant the originality that results from bold and independent thinking, but the originality that is sought for its own sake. The investigator who has no patience enough to listen till things themselves are going to speak in the situation he created; the thinker who cannot wait till time is ripe for judgement, who prematurely wants to construct and to systematize, pretending that nature or culture pressed this synthesis – they will come to conclusions that can hold up for some time until they are seen through in their objective weakness or fallacy, but that sooner or later turn out to have been if not erroneous, at least roundabout ways. It is true that courage to risk the danger of error belongs to the ethos of the man devoting himself to science, yet critical thoughtfulness should at any rate induce him to handing to his fellow-men the means by which they might check his view and separate truth from falsehood. The way, along which conclusions are reached, is more important than the assertion as such and this way must, in principle, be accessible to all.

Disinterestedness, courage and patience are, therefore, widening factors in scientific research, whereas impatience, fear and overgrowing service to interests have a checking and limiting effect.

But even the right ethical disposition being present with the investigator and thinker, other checks may turn up to the adequacy of knowledge to its object, e.g. when the *experiential basis* is too narrow, or the *formation of concepts* not imperative, or the *argumentation* not stringent, when the *system* can only unite part of the facts, when *wording* leads to misunderstanding.

First of all the insufficiency of the *experiential basis*. Not all facts and events occurring in the "world" are known, nor can they be made present. The domain of reality is vast and insurveyable, *spatial* distance can never be overcome entirely. Nor can all men of science together be present everywhere. More separating still is *temporal* distance veiling the future to us and making the past accessible only through its effects in the present and through our fading and transforming memory.

But science does not even want to observe all facts. Do not the *physical* sciences see in the individual facts "phenomena" i.e. *cases* in which general laws find their application and which are explained by the coincidence of these laws, occurring in the same way under the same circumstances? So through some exemplary facts they try to find the laws of these facts, which form a key to understand or predict again other facts. In the light of these theoretical objectives they will methodically look for quite definite experiences of facts and ignore an infinite series of other experiences as of no importance.

Never, however, will the theory designed cover the totality of experiences; this would only be so, when every empirical fact were *merely* a case: a univocal recurrence of exactly the same content. The domain, however, of what is repeatable in a homogeneous manner i.e. the domain of the purely numerical multiplication within uniform space and time, is an abstraction of the mind. And although mathematics suffer application pre-eminently in the so-called *exact* empirical sciences, it remains to be seen whether even there every fact is not somewhat a "single" event, that never recurs in exactly the same way under completely identical circumstances. It is true enough that the individual aspects are of minor importance there and that to master the future it will be sufficient for us to reach that degree of probability, which the abstract-general laws allow us. But the nearer we get to the higher regions of being- the spheres of *life*, of *psyche*, of *mind* – the less definitely imperative the character of the general laws becomes and the more we are to have an eye for the *individual* quality of events and beings.

However, in the sciences of *history* too completeness of the facts is not possible – and this is felt as a greater lacuna here than in the study of the more constant phenomena of nature. And even if the historian had plenty of documents concerning a particular period, he would out of an ocean of facts have to isolate the important ones in order to give his narration structure, shape and meaning. Who will tell him, however, what is of cultural importance and what is not? Here, too, fact and theory apparently interlock, since interpretation from pre-established connexions makes new facts important to history.

Moreover, it is a special difficulty of the sciences of history that the individual, historical fact lies somewhere in the past, where we were *not present*, whereas the physical fact as a general phenomenon can again and again be evoked by experiments and so can easily be checked by other people observing until it has been firmly established and stands like a rock in the surf. How are we by any chance to re-find historical facts "as they were", especially when these facts are not merely modifying the world of external objets, but above all the world of human relations?

It is the arduous task of the historian to trace out the mentality of persons and societies of long ago through the products of their mind as it has expressed itself in monuments and documents. This tracing can only be done by a constant, devoted and unselfish intimate intercourse with a very limited period; only then facts that are of importance and have a historical value and meaning will manifest their sense by the convergence of numerous indications. Isn't it necessary, therefore, that the historian stands *above his own mentality*, with its standards, judgements and appreciations (for this mentality is historically conditioned as well) in order to understand ideals and motives of a different directedness? At times he thinks he is succeeding in penetrating mentalities, that seem strange at first sight. Then again he finds he will never *completely* succeed, because, although he does have access to former cultures, the gate to them is the spirit of his own time. But still, this limitation of his science does not make him desperate, when he bears in mind that it belongs to the sense of human history to stick in our memory and there continue to play a part in the building up of our present and our future, so that at a certain temporal distance and as through a filter of what happened since, it becomes possible to interpret the past in its actualizable potentialities, and so in its value and meaning, which were not accessible to its contemporaries.

Does this include that the sciences that study the past are better off in a way than the sciences that observe *the actual social relations* in economics, laws, morals, language, art and religion, even though these draw on material more abundant and easier to access? To some extent, they are. For though a diagnosis of our times has the advantage for an observer to understand more easily the expressions of a civilization in the midst of which he himself stands, this connaturality need not necessarily be, it is true, yet easily threatens to be, an impediment to objectivity. For the fact, that the observer has a personal standpoint within the whole of society observed, causes the relations he discovers to be preferably the relations of the parts the others play towards the part which he plays himself. So that, in fact, not social life as it is in itself is observed, but as it appears to the observer from a definite situation within that life. This difficulty increases when he directly interferes by questioning his objects of study. The more so, when in stead of the questionnaire or the inquiry, there is the test by which the *psychologist* either enters with the testee into direct interchange of feelings, thoughts and intentions, or places him into an unnatural position of which the purpuse, be it vaguely, is also known to the testee.

And not only when we examine the psyche, but also when we study *life*, the object is placed in artificial circumstances in order to observe it more distinctly and accurately under a very particular aspect, but what we do not know with absolute certainty then, is: if it would behave exactly the same outside this field created by us.

Even in the physical sciences a certain revolution has taken place precisely by reflection on our possibilities of observing physical facts. Absolutely accurate observation was, of course, known to be impracticable as the act of measuring includes a physical interaction between the physical means of observation and the physical reality to be observed, and so the observation itself somewhat alters the situation. Quantum Mechanics, however, teaches us that, in principle, it is impossible exactly to determine at the same time both the velocity and the position of particles – so that we are never able to form an adequate and objective representation of facts by means of a corpuscular model.

When considering the insufficiency of the experiential basis and the limits of objective observation in all fields of empirical sciences, we at the same time pointed to the difficulties regarding *the exact formation of concepts;* for observing is indissolubly linked up with thinking. Each science uses concepts of its own and – even when borrowed from pre-scientific thinking – these concepts must be lifted to the level of that science by accurately pointing out their technical meaning according to the method of this science: the rise of the concept by proceeding experience must be demonstrated as necessary. So in physics we refer to experiment and instrument, to define what is to be understood by light, sound, heat, weight, mass and energy; in psychology we shall, according to the multitude of methods and theories, attach different shades of meanings to words like instinct, passion, sentiment, emotion, memory and intelligence; in the cultural sciences we have to refer to the convergence in a certain direction of the collected events, in order to get a wellfilled, charged, concept of a period comprised in a name like "Renaissance", "Baroque",

"Enlightenment". It is in the purification of concepts, their better adaptedness to grash the fulness of the data, that the spiral progress of sciences lies.

But forming concepts that are completely adapted to the phenomena is only succesfull *in very limited fields*, and only owing to that limitation. That is one of the reasons why the tree of sciences branches out more and more and that is also why each branch is going to lead increasingly independent life: the adaptation of the subject to the object, failing on a large scale, is more succesfully pursued on a small scale. The world as a whole continually slips from our grasp, it splits up into fields of nature, life, psyche, mind – whose interrelations remain obscure to us. To *specialize* in one of these regions – at the risk of going blind to the other ones – bears fruit and the more so as we confine ourselves again to a smaller particle within each field: "in der Beschränkung nur zeigt sich der Meister", might also be applied here.

The reverse side, however, of this fruitful specialization is *misunderstanding*. For each branch must, in order to express its appropriate concepts, form a language of its own. A purely artificial language, however, as introduced so profitably by mathematics and symbolic logic, cannot be applied alike in each of the sciences, because of the different experiential bases. That is why communication becomes increasingly difficult in those fields of scientific investigation where things are concerned more profound than the equally accessible obvious facts and their logical relations. In the cultural sciences confusion threatens to become greatest; especially the mathematician and the physicist will here feel the want of a clear, distinct and univocal expression. The ideal of universality and accessibility, of the many sciences conversing in one academic community, is indeed far from being realized.

With this limitation of intercourse due to our manner of forming concepts and expressing them is closely linked up another limitation: the lack of *argumentation stringent* and imperative to all. Every scientific exposition will naturally have to proceed logically, but that does not include that all trains of thought are stringent. If each of the spheres has got an objectivity of its own, then it has a sort of *probability* of its own too. We should not ask everywhere a certainty of the same kind as can be had in the most abstract field of mathematics. In things human a moral certainty can often be sufficient to make us convinced of the impossibility or at least improbability of the contrary. That is not the worst of it. For we might try to indicate exactly the degree of probability in propositions, interpretations or insights. That, however, is just what cannot be done alike everywhere. Although the naturel sciences are more succesful in keeping up the separation between well-established laws, verified theories and hypotheses to be verified, in the cultural sciences it is only by constant holding out in the developing process of science itself that the interpretations get their definite degree of certainty. So sober-mindedness and discretion in the way one presents one's views as correct is here an essential demand of sincerity.

So far we tried to shed light on limitations of the actual pursuit of science, which may never be looked upon as definite limits, because science in the process of its self-realization is constantly transcending them. The tension between the knowing subject and the object known gives rise to dynamic unrest and brisk life: science grows by purifying its ethos, by widening its experiential basis, by enhancing the objectivity of its observation, by specifying the meaning of words and concepts, by more stringent argumentation, by raising its degree of certainty.

The most fundamental question, however, to be dealt with now that we are coming to the end of our investigation, is the following: are there, apart from the limits arising from our lack of capacity, from the narrowness of our historical situation, from our necessity to arrange work by specialisation, are there still *more radical limits* that forbid to identify what is accessible to the positive sciences with being-as-such, which is accessible to man as man? *Is empirical science the only form of human knowing directed towards truth?* Or is it, that there are aspects of reality which may not quite absolutely escape empirical investigation, yet force the man of science to acknowledge that according to their real essence they lie in a dimension different from that of science, and accordingly would have to be reached, when we men can in fact reach them at all, by attitudes and methods different from those of science?

That every specific science by its specific nature acknowledges, besides its own, other fields or other methods of research is obvious, though here out of the question. For the question has been narrowed down to *the sciences as a whole* – even though this whole is not systematizable – and thought of as *ideally accomplished*. The answer, therefore, to this question is not to be found within the specific sciences, but can only come from the scientist as man, wondering where in the relation between man and world, the scientific impulse arises. Such a reflection on the *foundation* of science may well originate from the pursuit of science – when the physicist experiences, as a problem, the attractive force of mathematics and formal logic, the historian his connection with the moral sciences, with their standards and metaphysics. But this reflection *transcends* all the same *the empirical way of approach*, as the abstraction, by which every specific degree of science originates in its proper self, is abandoned in order to turn to the concrete starting-point of knowledge. For, the reason why the scientifically constructed whole of objective science cannot cover being-as-a-whole is that science itself must of necessity have its *origin* genetically *in pre-suppositions*, which, it is true, it cannot deny, but cannot verify scientifically either.

There is first of all as a pre-supposition *the given world-objects as such*, insurveyable as a whole and beyond our reach both as to spatio-temporal and as to qualitative completeness. Within that world science does establish facts and rationally interconnects these facts. But these general connections are merely hypothetical necessities, whose comprehensibility does not make comprehensible the very *existence* of the individual-concrete beings. Not, that it would be allowed to call reality-as-a-whole, because of this striking the bottom of facticity, irrational – i.e. opposed to reason–. Anyway, for scientific thinking facticity remains a starting-point, that ought to be accepted as a limit beyond which there is no re-questioning.

Notwithstanding this, the question does arise, *what the sense and the meaning is* of this world-as-a-whole, that appears to us as an original datum and in which the sciences isolate objects of reseach and specific investigation. And how is it, that *we* have been *given* this world? After which another question immediately crops up: *to whom* has it been given after all, *who is the subject?*

When we restrict ourselves to science as relation between

subject and object, the world presents itself to the subject that pursues science, i.e. to *common reason with its structures* of observation and thought. But these structures, together with the given world, belong to the pre-suppositions of empirical science. They, therefore, cannot in their turn be investigated by objective science of facts, no more can the logical and methodological principles flowing from them. Sociology of science is no doubt an important branch of science, which teaches us how to understand psychologically and sociologically scientific thinking in its historical mode of becoming. But nevertheless it cannot make pronouncements on the very *validity* of scientific thinking, on its worth of truth. This would require a doctrine which took scientific thinking not as an observable *object*, but in its originating from *subjectivity*. And so, science points to a basic examination that transcends science itself. Such a basic examination may be called a philosophy of knowing that tries to elucidate knowing as such, no longer considering it as a clear-cut object among objects, but as the bond itself between subject and object.

The question, however, must be asked, whether the world presents itself to the subject only in so far as the subject pursues just science. From *vulgar* experience the man of science has it, that he is more than man of science and not the only man existing. He thinks himself to be in a *world of persons*, who behave as conscious and free people, having intercourse with one another in feeling, thinking and acting. In psychology and sociology, in history and in other sciences of culture, this personal and intersubjective atmosphere is considered in its observable manifestations: in behaviour and work. In those sciences it becomes clear that the human mind is directed towards truth non only in science, but also in technique and economics, arts and letters, ethics and religion. These different fields are described and analysed, also in their specific directedness to truth and value. But empirical science lacks methods to gain access *to the human mind in its subjectivity as such*, in its self-consciousness, freedom and responsability. It may e.g. to some extent approach the artistic consciousness by way of the manifestation of it, but it does not experience reality the way the artist does in his artistic consciousness. The same applies to the ethical and religious attitude of mind. No more can the quest for the fundamental relation be-

tween the aesthetic, the intellectual and the ethical consciousness, i.e. for the common root of feeling, knowing and willing, of
thinking and acting, be answered empirically. And whether the
empirical multiplicity of subjects of consciousness proceeds from
one "transcendental subject", yes or no, or, if not, how the communication of minds is to be understood, these are questions to
which observable facts are little likely to give the answer. And
although positive science does not fail to realize that there *is*
connection between nature and culture, between the sferes of life,
psyche and mind, yet how we are to understand this connection
ultimately, is a matter of philosophical reflection and lies beyond
experiment.

What ultimately the sense and meaning is of man's existence
in time, what he has to hope for and how he has to live, those are
ultimate questions which cannot be solved by referring to data
indisputably pre-established already, because they stake everything at once – *the totality of subject and object* and the *cause* of
their bond of union. The ultimate cause by which whatever is,
ultimately is what it is, the last ground of being, the Absolute, to
which anything relative must needs relate, that ultimate cause
transcends positive-scientific investigation of facts. When we,
human people, do have access to it, then it is not because we turn
our look on it as on an object lying before and outside us, but only
because we are allowed to participate in a degree in the act by
which the Absolute constitutes itself in pure actuality and self-
lucidity.

Positive scientific knowledge, therefore, is neither the start nor
the finish of our knowledge. We were not born and will not die
"as" scientific investigators. Science originates from somewhere
in the totality of man and society, and is determined and finited
by that *situation*. It comes about and grows in a multiplicity of
particular investigations concerning particular fields. Although
the fields of investigation extend further and further, although
the results are more and more important and fruitful to life, yet
ultimately it is not science, that determines how we are to live,
since the horizon of our life is limitless and infinite and is beyond
all knowledge of facts. Conscious of its limits, science in itself
points to what might be attained *through other attitudes* e.g. those

of the philosopher, but as much those of the creating artist, of the man with moral sense, or the man of religious faith.

And yet, science in the man pursuing it remains allied to these other attitudes. Science preserves in particular the bond of union with *philosophy*, which also proceeds "in the manner of science". The fact that there remains a distinction and how (in spite of this distinction) there can be a connection between the methods of empirical science and the though different yet non unscientific methods of philosophy in epistemology, in philosophy of nature and culture, in anthropology and ontology, those are problems which do not concern us here. For it is not our task to explore whether and how in a radical criticism philosophy might test the ontological and logical pre-suppositions of the specific sciences, as well as of the pre-scientific "philosophies of life".

For the man of science it suffices to know, that by acknowledging the essential limits of the object attainable through empirical science intercourse is also possible with those dissenting from him with respect to the ultimate questions of life. To perform the abstraction which inaugurates science, need not mean to him to sever completely the connection with this outlook on life. For, both the start of positive-scientific investigation and the choice of its object often will spring from motives beyond science itself and they, no doubt, will determine the climate of the investigation; every man, including the one who pretends to have only an eye for his specific phenomenon, projects a provisional all-embracing world-picture, which never springs from the mere data themselves. But what must be prevented is: that this super-scientific stimulus determines *the content itself* of scientific establishments. The individual investigator is to deposit his findings, conclusions, opinions and insights before the forum of the community of investigators to have them corrected, approved or disapproved.

He who lives consciously and willingly from inspirations deeper than positive science can reveal, will with perfect confidence enter into this communication with those dissenting from him. He knows, that every particle of truth has an origin higher than human, even the limited truths which at an increasing speed science detects, – thereby making us delighted and anxious, delighted in the discovery, anxious about the responsability.

ON THE BOUNDARIES OF SCIENCE

by R. KRONIG

Your Majesty, Ladies and Gentlemen:

The task with which I have been entrusted, namely to speak about the boundaries of the natural sciences, brings with it the consideration of the purpose of these sciences.

The *pure natural sciences* attempt to give a description of the observations we make with our sense organs. In carrying out such observations we are assisted to a large degree by various instruments of observation and measurement. When discussing the boundaries of the natural sciences in general we shall then have to direct our attention towards the limits to which our techniques of observation and measurement are subject. Every improvement of old and every development of new methods of observation and measurement will be immediately followed by a displacement of the boundaries of what is accessible to natural science. A number of examples from the history of natural science may serve to illustrate this.

As a first example we mention here the invention of the telescope and the progressive improvement of this instrument. Already the application of the simplest telescopes by Galileo and Huygens led to the discovery of new astronomical objects. By increasing the size of the telescopes, by refining the optical parts and by using photographic instead of visual registration a constantly increasing part of the cosmos could be subjected to astronomical study. The discovery of spectral analysis by Bunsen and Kirchhoff in the last century opened up the possibility of supplementing the localization of astronomical objects by data concerning their velocity and chemical composition. During the last ten years the armoury of the astronomers was enlarged with completely new accessories, namely the instruments of radio-astronomy, radar mirrors and their electronic equipment. With their help it became possible to penetrate into regions of space, where by reason of the absorption of dust or gas clouds the eye, aided by ordinary telescopes, is useless. Not only was it possible in this way to discover new stars, the radio-stars, but our countrymen Oort and van de Hulst also succeeded in demonstrating that our own galactic system, just like so many others spread throughout the universe, has a spiral structure.

If we remain nearer home, then we should mention the great extension of our knowledge of the atmosphere made recently by high altitude rockets. The behaviour of pressure and temperature and of the chemical composition of the atmosphere as a function of altitude have thereby ceased to be an object of scientific speculation and have become accessible to quantitative investigation. Also we are to-day much better informed about the spectral composition of the light of the sun at great heights, which partially suffers strong absorption when penetrating to the surface of the earth.

For our understanding of the interior of the earth a knowledge of the behaviour of matter at extremely high pressures is of eminent importance. The work of Bridgman in America, who was able in his laboratory to attain pressures of the order of 100 000 atmospheres and also to carry out measurements with them, should therefore be of high interest to the geophysicist. However, even to-day little is known of the earth's interior, and geomagnetism, first noticed many centuries ago, is still very imperfectly understood.

The displacement of the limits for another variable, which next to pressure characterizes macroscopic matter, viz. the temperature, is an exciting story in which the Netherlands have played an honourable rôle. The attainment of extremely low temperatures has proved to be important especially for physics itself. A lowering of temperature signifies, indeed, the counteracting of the irregular motions of the atoms and molecules of which matter is composed and in conjunction with it an increased order in the structure of matter. Every step in the direction of the so-called absolute zero has led to the discovery of new phenomena. I mention only, keeping near home, the discovery of superconduction by Kamerlingh Onnes, of the miraculous liquid helium II by Keesom and of the de Haas-van Alphen effect. At the other extreme, that of very high temperatures, we are chiefly restricted to what nature presents us in the stars, although on the earth too it has become possible to produce astronomical temperatures by the explosion of atomic bombs.

While the development of the telescope was of particular

importance for astronomy, another optical instrument, the microscope, has been of decisive influence on biology. Here too already the simplest instruments in the hands of Antonie van Leeuwenhoek furnished a rich harvest of new results. The discovery of micro-organisms, of the blood corpuscles and of the germ-cells, together with that of the cell-structure of higher plants and animals, has opened unprecedented perspectives. While in the beginning an improvement of the microscope was striven after, it soon became clear that the resolving power of this instrument is limited by the finite wave length of light. Nevertheless in later years striking innovations of an instrumental nature have been introduced in the microscope, of which I only mention Zernike's phase contrast method which enables us to make differences in index of refraction visible as differences of intensity. Another important step was the development of the electron microscope, which allows us to increase the resolving power by a factor 100. By means of this instrument a number of assumptions could be demonstrated *ad oculos* which had been proposed by biologists as pure conjectures. Making visible the giant molecules of organic chemistry on the one hand, of viruses on the other, is an important step in bridging the gap between inanimate and living nature. That at present attempts are being made, e.g. in France, to construct a proton microscope, gives us reason to hope that in the field of microscopy too the last word has not yet been spoken.

We further wish to mention a subject in which a radical enlargement of boundaries has also taken place during the last fifty years; I refer to the experiments with fast charged particles. At the beginning of this century cathode and positive ray tubes were known in which electrons and ions respectively could be accelerated by tensions of the order of 100 000 volts. At present the betatron, cyclotron and proton synchrotron allow us to give to these particles velocities which correspond to voltage differences of several thousand million volts. Such fast particles can serve on the one hand to disintegrate atomic nuclei and to realize thereby the age-old dream of the alchemists, the transformation of the chemical elements; on the other hand they furnish the opportunity to imitate in the laboratory the phenomena which take place in the atmosphere under the influence of cosmic radiation.

Indeed, cosmic radiation consists of exactly such extremely fast particles which, coming from stellar space, constantly bombard our earth.

Until now our attention was principally directed towards the displacement of boundaries in the natural sciences, due to the improvement and development of instrumental techniques. Just as important, however, for this displacement has been the increased control in the making of new substances and materials.

Considering first the inorganic sciences, we cannot pass over the great progress achieved in the domain of ferromagnetism by the preparation of new metallic alloys and non-metallic compounds. Also the rapid increase in our knowledge of semi-conductors deserves mention, which is principally owing to methods of preparing the substances in question with a hitherto unknown degree of purity.

Even more striking, however, are investigations in which biology and organic chemistry have closely cooperated. The discovery of substances, such as the sexual hormones and the vitamins, which produce important effects in the living organism in extremely small quantities, has been the result. Not only has it been possible to specify and isolate these substances in organisms, but several of them could be prepared synthetically in the chemical laboratory. In view of the extremely small quantities of substance with which one often has to deal, the application of completely new methods of chemical micro-analysis have been indispensable in this connection.

Finally we must mention two domains of investigation in which new methods of research have also opened surprising perspectives. The use of radio-active indicators, i.e. of atoms which signalize their presence by radio-activity, has made it feasible to follow up in detail the metabolism of living organisms. By the use of X-rays and radio-active rays, which can bring about changes in the parts of the cellular nucleus that are decisive for heredity, the chromosomes, it became possible to give a more direct basis to various hypotheses of the theory of heredity.

Until now our attention has been directed toward observations in the natural sciences and the limits imposed by experimental technique. Natural science, however, is more than the collecting of observations. Besides its empirical aspects it also has its theoretical side: the ordering of the observations with the help of concepts and the formulation of general statements concerning these observations in the form of hypotheses and natural laws.

Just like the collecting of observations the obtaining of such general insight is a gradual and difficult process. While in the former the resistance of matter makes itself felt, which the experimenter has to overcome with infinite patience, in the latter the theoretician must transcend psychological resistances in the form of traditions and habits of thought. A new insight therefore in general first makes headway in intuitive form. It emerges as an image from the subconscious, from what G o e t h e has called "the realm of the mothers", taking on henceforth more conscious, and finally abstract shape. The theoretical ideas thus have their origin in certain structural properties of the human psyche, best denoted by the term "archetypes", which goes back to S t. A u g u s t i n e.

This is perhaps the suitable moment in this lecture to point out that the dualism in the practice of the natural sciences just mentioned is a fundamental one. That we do not regard our sense reactions as purely mental manifestations of a primary nature, but postulate behind them an external world in space and time is principally due to the circumstance that these reactions have aspects entirely independent of our own will; aspects which make it possible to repeat experiments with identical result. That on the other hand it is naive to consider the theoretical insight into the events of nature and the constructions of which it makes use as a sort of epiphenomenon of these events, suggested by the sense reactions and derived from them, can hardly be stated with sufficient emphasis in our extraverted civilisation. When, however, we remember how truly revolutionary ideas come to us, spontaneously entering our consciousness provided we have adequately prepared it for their reception, then we will not disparage the rôle of the human soul in every natural science as against that of the sense data.

The strong satisfaction which the attainment of a fundamental scientific insight gives, undoubtedly has its roots in a pre-stabilized harmony between the internal world of archetypes and the external world of facts. Again and again theoreticians have felt this as a great wonder. It is interesting to mention what K e p l e r, this important figure at the threshold of modern rationalistic science, says on this subject [1]:

"Understanding signifies bringing what is observed outside together with the internal ideas and judging their correspondence; what Proclus has expressed very beautifully with the word "to awake" as from sleep. Indeed, just as what we encounter outside reminds us of what we knew before, thus the sense observations, if they are recognized, call forth the intellectual and internally present ideas so that they become visible to the soul, while before they were hidden there in veiled form. How have they come there? All ideas and forms of which I just spoke are present in the beings which have the capacity of knowledge; they are by no means assimilated discursively, but depend on a natural instinct and are innate, as e.g. the number of petals is innate in a plant or the number of chambers in the apple" [2].

In our days Einstein has expressed himself in similar terms regarding the miracle in the correspondence of our forms of thinking and the observations in the external world.

When we now inquire into the history of the concepts of the natural sciences in their concrete aspects, then of course in this lecture we can only touch upon the most striking cases where human thought freed itself from ancient prejudices and thereby extended the boundaries to which it was subject.

The ancient Greeks already put forward the idea of the atomistic structure of matter, which attempts to explain the multiplicity of phenomena with a limited number of essentially different elementary particles. When applied to inanimate matter this is to-day the backbone of physics and chemistry. Supported by the work of Lavoisier, Dalton at the end of the

---

[1] Reference is made in this connection to the article by W. Pauli in Jung-Pauli, "Naturerklarung und Psyche", Rascher Verlag, Zurich, 1952.

[2] J Kepler, Harmonices mundi, book IV; ed. Frisch, bd. V, p. 224.

"Nam agnoscere est, externum sensile cum ideis internis conferre eisque congruum judicare. Quod pulchre exprimit Proclus vocabulo suscitandi, velut e somno. Sicut enim sensilia foris occurentia faciunt nos recordari eorum, quae antea cognoveramus, sic mathemata sensilia, si agnoscuntur, eliciunt igitur intellectualia ante intus praesentia, ut nunc actu reluceant in anima, quae prius veluti sub velo potentiae latebant in ea. Quomodo igitur irruperunt intro? Respondeo, omnino ideas seu formales rationes harmonicarum, ut de iis supra disserebamus, inesse iis, quae hac agnoscendi facultate pollent, sed non demum introrsum recipi per discursum, quin potius ex instinctu naturali dependere iisque connasci, ut formis plantarum connascitur numerus foliorum in flore et cellularum in pomo."

eighteenth and the beginning of the nineteenth century developed the concepts of chemical elements and compounds. In our days, where the atomic nuclei no longer are regarded as immutable but themselves as composite, a nuclear chemistry has evolved next to ordinary chemistry, having as basis the same idea of elementary particles, in this case the protons and neutrons. The realization that the number of constellations which can be obtained from a small number of different elements by combination and permutation is enormously large has also played a rôle in biology, where it offers the possibility of considering the hereditary qualities as localized in the chromosomes.

That the elementary particles of matter are in constant motion is a very old idea too. But only in the physics of the last century was this notion given a more definite form by connecting the motion of the atoms with the concept of temperature. The success of this so-called kinetic theory of matter has given the incentive to attempts at reducing all phenomena to mechanical processes. This mechanistic picture of the world, which actually goes back to Descartes, has had consequences far outside the range of physics proper. In its light one was inclined to ascribe to the world a strictly deterministic character and to designate the experience of a free will and of a moral consciousness deriving from it as illusory.

Through the development of scientific thought, however, these views have long been surpassed. In the first place the study of electric and magnetic phenomena by Faraday led to the introduction into physics of the field concept, which found its mathematical expression in the hands of Maxwell. In addition the concept of energy displaced more and more the visual concept of motion. To this the discovery of the law of conservation of energy contributed essentially, taking its place next to the older law of the conservation of matter.

The tendency of the theory to assume more abstract forms manifested itself particularly in the development of Einstein's special theory of relativity. In this theory the necessity was realized of subjecting the concepts of space and time to a close analysis. It appeared then that in measurements of time and distance the choice of the coordinate system in which they are carried out is relevant. A further consequence of this theory was that the laws of conservation of matter and energy were fused into one by reason of the discovery that every change of energy of a system is accompanied by a change in mass. In modern nuclear physics, where we are used to the idea of material particles being created or annihilated under absorption or emission of radiation, this view has found direct experimental confirmation.

The general theory of relativity goes still one step further in the removal of boundaries inherent in conventional thought. By ascribing a non-Euclidean geometry to physical space, which, however, manifests itself only over cosmic distances, it became possible to incorporate gravitational phenomena into modern physics in a satisfactory way. This led to extensive new perspectives in astronomy and cosmology. In particular the circumstance that a non-Euclidean space even in the absence of boundaries can have a finite volume gave rise to important considerations. The theories of the expanding universe, which are supported by spectroscopic evidence on spiral nebulae, in addition wish to attribute a finite lifetime to the cosmos, at least as far as the past is concerned. That the values for this lifetime resulting from the astronomical data lie close to the values found for the age of the earth on the basis of geology is extremely suggestive.

An entirely new situation was created in natural science by the arrival of the quantum theory. Here the idea was imposed that one of the primary assumptions of classical physics cannot be maintained; viz. that it should be possible by proper care to make the influence of the observer on the object observed as small as one pleases. As the experimental facts demonstrated, nature is so constituted that subject and object of an observation can never in principle be separated sharply. In theory this feature finds its expression in the wave mechanics of de B r o g l i e and the uncertainty relations of H e i s e n b e r g, resulting from it. It is true that the wave function which describes a mechanical system in this theory behaves deterministically; its form at a given moment follows from that at a previous moment by means of a differential equation, the wave equation of S c h r ö d i n g e r or D i r a c. But the connection between this wave function and the results of measurements, e.g. the appearance of particles constituting the system at definite points of space, has a statistical character, as was first realized by B o r n. This statistical character, according to our present knowledge, is inherent in the phenomena themselves and not the result of an inadequate technique of observation or measurement.

The evolution just described is of primary importance, also far outside physics itself. In the first place it has erased the boundaries between theoretical physics and theoretical chemisstry. Indeed, quantum mechanics offers the possibility of regarding typically chemical concepts, like valency and affinity, as consequences of the elementary electromagnetic interactions between the building stones of matter.

As regards in particular the impossibility of sharply separating subject and object of an observation, this undoubtedly is of even greater significance for biology than for the inorganic sciences. As has been stressed repeatedly by B o h r, every attempt to gather data about a living organism has far-reaching consequences for this organism; so much that a too radical interference will give rise to the death of the organism, which then will disintegrate according to the laws of chemistry.

Similar remarks apply to experimental psychology and neurology. Thus the psycho-analyst and his patient form a unit, the success of the treatment depending essentially on the willingness of the practitioner to abandon the rôle of the pure observer and to make conscious the complexes of his patient by his participation, called "transfer" by F r e u d.

Finally, a change in point of view in natural science, such as took place in passing from classical to quantum physics, cannot fail to have consequences for our general philosophy. Indeed, whatever occurs in the mind of thinkers and poets, manifests itself one or two generations later in the events of every day. As illustrative of this assertion I would like to call your attention to what H e i n r i c h H e i n e has written on the subject more than a hundred years ago in his essay "Geschichte der Philosophie und Religion in Deutschland". Conceptions of life such as the dogma of predestination in Calvinism or that of deterministic economics with its eschatology of Marxism are unquestionably related in a subtle manner to the deterministic aspect of classical physics.

Until now I have tried to give you a kaleidoscopical review of

4

the ideas that played a dominant rôle in natural science in the more distant as well as the recent past. You will justly ask what are the great problems to-day and in which directions fundamental developments may be expected in the near future. That predictions in this connection should be made with the necessary reserve is self-evident.

In physics it is clearly a problem to account for the existence of the elementary particles of which matter is composed, as well as for their properties. Also a harmonious unification of the theory of relativity on the one hand and of the quantum theory on the other is highly desirable. Probably further modifications in our notions concerning space and time will be necessary in this connection. That the theoreticians have hardly made any progress in this complexity of problems during the last 25 years has undoubtedly its origin in the circumstance that again deeply rooted habits of thought must be abandoned.

In the biological field a further penetration into the characteristics that distinguish living and inanimate matter is perhaps the most urgent task. Biophysics and biochemistry will play an important rôle in its solution. In particular the study of phenomena in the living nerve is promising. Attention has been drawn repeatedly during the last few years to the interesting parallelism between the action of the brain and the functioning of modern automatic telephone systems and computing machines.

In the field of psychology, finally, the investigations carried out by Rhine deserve serious attention. His experiments have reference to cards provided with various simple figures, which are submitted to the experimental subject with the illustrated side downwards. The subject must then state which figure is present on the invisible side. According to the experiments the number of hits in general is larger, and sometimes considerably larger, than would be expected from the laws of chance. In view of the size of these deviations and the very respectable number of experiments it is extremely improbable that the more than expected hits are a result of fluctuations.

In case a further elaboration of the methods of Rhine and his successors and a delimitation of the factors and circumstances involved should continue to prove the reality of the phenomenon, which has been denoted by Rhine as "extra-sensory perception", this would be of the most extreme importance. From different sides, e.g. by Jung, it has been suggested that the infringement of a description in space and time manifesting itself therein has something to do with the coming into operation of the unconscious part of the human psyche. However this may be, here unquestionably lies a field for joint investigation by the physicist and the psychologist.

Until now our attention has been directed exclusively towards the pure natural sciences. The satisfaction which their study gives may perhaps in its highest forms be qualified as religious; at any rate it rests on the search of the human mind for truth and the desire to express the knowledge acquired in harmonious forms, giving simultaneously an intellectual and aesthetic satisfaction. Just like the fine arts pure science concerns man essentially as an individual. In the case of the *applied natural sciences*, on the other hand, also the cultural aspect must be touched upon, for they have often had a radical influence on human society. When inquiring for boundaries here, we will have to look at the problem also from this point of view. In consequence the accent will fall more on the situation of our time than in the first part.

In the applied natural sciences, which substantially may be divided into the technical, the agrarian and the medical sciences, we encounter in contrast to the pure sciences the new factor of utility or at least that of the attainment of certain practical objectives. The application of our insight into surrounding nature is as old as humanity itself. The conscious use of tools, indeed, next to a language employing concepts and the ability to laugh, is one of the most fundamental characteristics which distinguish man from the animals. We encounter tools already in the stone age in the form of stone knives, but also in the form of stone weapons, i.e. both for constructive and destructive purposes; a dualism which goes through the whole of human history. The plough is probably just as old, and unquestionably also the use of medicines, prepared from roots or plants, goes back to prehistoric times, although naturally no traces have been preserved.

Next to tools we meet as a second important factor, at least in the technical and agrarian applications, already in very ancient times the use of sources of energy other than that of human muscular force; in the first place the muscular force of animals; next to it the propelling force of water and wind in mills and ships; finally the caloric energy, freed during combustion in fire and employed for purposes of heating, but also in the chemical processes in the kitchen and in the preparation of metals from their ores. In the use of tools, energy sources and medicines, a continuous thread runs through human history, and although the distance between the resources of modern man and of his predecessor in the stone age is very large, one can only speak of a quantitative, and not of a qualitative difference.

The question has often been asked what status technical inventions should be given and if they may be named in one breath with the creations of religion, art, or pure science. It seems to me that the technical activity of man forms a category by itself in which the directive factor is a practical purpose, and that it takes a fundamental place among human creative occupations. The joy which the born inventor experiences from his invention surely is quite commensurable with the satisfaction rewarding the man of science or the artist.

If we now ask for boundaries in the applied natural sciences, then we must remember in the first place that a *de facto* limitation is always imposed by the status of pure science. Indeed, only knowledge already attained can be applied. It should be remarked, however, that the interdependence of pure and applied natural science has not a one-sided character. For many pure scientific investigations have had their origin in practical problems, and in addition, as already mentioned at the beginning of this lecture, progress in any experimental science is intimately connected with the construction of certain instruments, requiring a sufficient measure of technical proficiency. We can thus justly speak of a close interaction between pure and applied natural science, such that the factual limits of the one entail limits for the other one too.

In the second place, however, the question arises whether limitations should be imposed on the application of the natural sciences in order that the cultural life of humanity may not be endangered.

We shall first discuss those applications and inventions which seem in themselves to have a constructive, or at least a neutral character. As mentioned before, man in general cannot be considered apart from his technical activity. However, in particular in our Western civilisation developments have taken place since the end of the 18th century, which in connection with the question just posed deserve our full attention. I refer to the process, so fittingly denoted by the Anglo-Saxon term "industrial revolution", which phenomenon has made the influence of the West felt over the entire globe and which would have been impossible without a large measure of technical skill. Yet it should be understood that it has its ultimate incentives in certain psychological attributes of modern Western man, rather than in this technical skill itself.

Among these attributes should be mentioned above all the urge to organize, which was still essentially strange to the Middle Ages, the Renaissance and the period of enlightenment, and first manifested itself clearly with the coming of N a p o l e o n. It is illustrative for what was said before that this quality in the beginning did not show itself in the technical field, but in the organisation of the state and in military matters in the form of compulsory service in contrast to the armies of hired soldiers previously in use.

In the preceding centuries the relationship between man and his fellows had been almost exclusively one of personal contact: of the members of a family with each other; of the peasant with the inhabitants of his village or with the merchant; of the artisan with his journeymen and apprentices etc. Government too had a personal form in the sovereign, whom one could see, pay homage to – or assassinate in case he made too much abuse of his power. Advantageously the terms "action at a distance" and "action at close range" can be taken over from physics in this connection. Human relationships before 1800 had essentially the character of close range actions. This remark even applies to the all-embracing institution of the church which brought forward the personal element in the contact between the priest and the members of his parish.

Under N a p o l e o n all this becomes definitely different. Concrete ties are gradually replaced by abstract constraints, organically grown relationships by artificial constructions; general rules, intellectually conceived, take the place of old traditions and customs. I refer e.g. to the subdivision of France into departments that had nothing to do with the historical past, and to the institution of the registrar's office and general military conscription.

All these phenomena do not yet concern the application of technical inventions. Only later, in the first half of the 19th century, the new spirit begins to dominate the field of engineering. Here it is production in the factory in which it finds expression. As already mentioned, neither the use of tools, nor that of non-human energy, nor the combination of these two in the machine is of recent date. Nor was the cooperation of a smaller or larger number of persons unknown in tasks which transcend the power of the individual. In the building of the Gothic cathedrals many artisans were employed, just as e.g. in the construction of the already quite large ships that crossed the oceans during the mercantile period.

A new feature in the factory labour of the 19th century is on the one hand the fact that the owner of the means of production no longer is identical with the persons who carry out the production. An additional new feature is the extreme division of labour in which the workers perform only a few, often very simple manipulations so that their personal relation to the final product becomes more or less lost. It also deserves notice that this division of labour does not have its origin primarily in the limited powers of the individual or in the circumstance that he can be an expert only in a limited field, but in a point of view which suddenly comes to the fore more or less consciously everywhere, namely that of industrial efficiency. Human labour is thereby deprived of its individual, personal character and enters more and more into an anonymous sphere.

The growth of these aspects in human society finds expression in the second half of the 19th century in a new legal concept, that of the limited liability company, appropriately called in French "société anonyme". The tripartition of industry into owner-shareholders, management and labourers resulting therefrom, together with the constantly increasing size of industrial enterprises, accentuated this anonymity still further.

The material abuses which in the beginning were the consequence of this development, such as child-labour, excessive working hours and danger of accidents in the factories, gave rise to the establishment of labour unions, to which the term "société anonyme" too can justly be applied in view of the fact that their members individually also disappear completely in the large quantity. These unions, supported by some of the more far-sighted among the employers, succeeded in the long run in checking the abuses just mentioned. The resulting reforms, however, were chiefly confined to the material sphere, while the psychological aspects which are closely related to the method of production and the structure of industry were hardly taken into account. By transforming industrial enterprises into a kind of foundations in favour of all persons from high to low connected with them, taking into account their risks and achievements, it might have been possible in principle to restore at least to some degree the personal relation between man and his work, without entirely giving up industrialization.

Instead of repressing as much as possible the anonymous elements in human society, socialism came forward with the idea of carrying this tendency to extremes by giving the last word to the most abstract impersonal factor in society viz. the state. Since then, however, also in countries in which socialism in spite of strong labour unions never played an important official rôle in the political system, as e.g. in the United States, the power of the state has increased to an alarming degree since the beginning of our century. G o e t h e, who foresaw this development like so many other things, pictures it already in the last act of the second part of "Faust", where the protagonist, who in the first part strove with all the fibres of his being after knowledge of life, becomes the victim of a purely organisatorial activity and like a M u s s o l i n i performs reclamation of land on a large scale. N i e t z s c h e too foresaw such happenings at an early date and in "Also sprach Zarathustra" devotes a word to the state, "the cold monster".

If we now return to our starting point, viz. the question to what extent the progress of applied science as such can be held responsible for this development, then I think the answer must be negative. The misfortune that our Western society is more and more taking on an impersonal and regulated character must rather be ascribed to the extraverted attitude and the preponderance of the element of volition in the entire Western world. This world has gradually become dominated by the fetish of efficiency and the idol of general welfare and social security; not keeping in mind the words of Shakespeare in Macbeth: "Security is mortals' chiefest enemy". Naturally man has always given, and will in all future have to give, much care and attention to his material existence. This, however, does not justify the status which this theme increasingly occupies in the thought of modern man at the cost of spiritual values.

It is an obvious question to ask what has been gained at the price of the collectivisation of large sectors of our life in the form of industrial world corporations, welfare states, people's democracies, etc.; and it is perhaps instructive in this connection to compare in concrete terms our life of to-day with that of our ancestors, e.g. in the 17th century. We may remind ourselves in this connection that many of us had the opportunity in the winter 1944–1945 to study the effect on our daily life of the absence of a number of modern inventions.

Considering first the necessities of life, to begin with our housing, it must be admitted directly that in this field not so very much has changed and that our ancestors constructed buildings which to-day still defy time by their solidity, in which they differ favourably from many buildings of the period after the second world-war. Sanitary equipment, running water, gas and electric lighting, on the other hand, are acquisitions of which all of us will gladly testify that we would be very unwilling to do without them.

In the field of clothing we have at our disposal to-day fabrics made of artificial fibres by the side of wool and linen fabrics, known already in olden times. Although our women-folk may value nylon stockings very highly, it can hardly be maintained that they should be counted among the necessities of life.

A careful study of the cookery-books of the past teaches us that the dishes composing our menu to-day are not so very different from those of a few centuries ago. The development of the beet-sugar industry has turned sugar from an article of luxury to one for popular consumption. Undoubtedly many dishes have thus become more attractive, but just as undoubtedly dental caries has become more frequent thereby.

We see thus, on the one hand, that the changes of what may be included in the necessities of life, brought about by the application of science, are not so very radical. In the field of medicine on the other hand, in particular that of surgery and pharmacology, science has borne rich fruits. The liberation of man from bodily pain and bodily defects, indeed, is an entirely positive result which would have been impossible without scientific knowledge.

Many other applications of science, however, should be judged with a great deal of reserve. They have often created new needs and desires which have little to do with true joy of life and deeper humanity and which often even are harmful in this respect. That is particularly true of the development of methods of rapid transportation and communication. Also regarding the effect of these Nietzsche in "Also sprach Zarathustra" has given a pessimistic prognosis: "Die Erde ist dann klein geworden und auf ihr hüpft der letzte Mensch, der alles klein macht". Not only do they bring about that by the dispersion of good friends and members of a family over the whole globe valuable emotional ties are slackened; the agressive character of western civilisation has also been given the opportunity through these inventions to exert a strongly disintegrating influence in regions where cultures, partly of much older date, had evolved well-defined forms of life. Rarely has a more impudent word of self-overestimation been spoken than that of "the white man's burden". Now that colonial relationships everywhere are coming to an end, the attempt is made here and there to compensate the diminishing of western influence in the political sphere under the guise of "aid to under-developed areas".

There is still another large group of problems caused by the application of science, namely the growth of the world population which has been made possible by it and which has been promoted by the progress of medical science. This purely quantitative result cannot be counted as other than negative. Together with the diminution of isolation due to the development of methods of rapid transport and communication, it brings about tensions and their discharge in catastrophic collective actions which formerly would have been quite unthinkable. A conscious limitation and gradual reduction of the world population seems a first condition for avoiding serious explosions in the future and for maintaining the possibility of cultural life.

Our attention has hitherto been directed to dangers to our civilisation arising from applications of science which at first sight appear innocuous. A word must still be said about the intentional use of scientific knowledge for destructive purposes, in particular those of war. Until recently it was an exception that pure scientists were willing to take part in such activities. At present this unfortunately cannot be said any more. Thus the development of the atomic bomb would have been quite impossible unless a number of physicists and chemists, among them quite prominent ones, had given their aid to this sinister project. Further examples in this category are applications of bacteriology and psychology for militairy purposes, of which the latter are perhaps the most dangerous. Characteristic for all these and similar applications is that they can serve for the strengthening of power of smaller or larger groups of men, so that it is appropriate to speak here of a prostitution of science.

In an interesting book, entitled "The next million years" Sir Charles Darwin, grandson of the propounder of natural selection, has attempted to give a vision of the long range future of humanity. While the material factors have carefully been analyzed, it seems to me that Darwin has taken the factor of the human mind too little into account. Toynbee, in "A study of History", goes much deeper in this respect by investigating under what circumstances civilisations can change their direction or even disintegrate. In the light of history, indeed, it is not at all unthinkable that already in the nearer future, i.e. in the course of a few hundred years, the extraversion of modern Western humanity may give place to an introversion, such as happened for instance during the decline of the Roman Empire in the period of early Christianity. If as a consequence the interest of creative individuals turns away from the outside world, then this can only result in technical procedures becoming much more primitive. To take a view into our own future is not given us, but nothing is perhaps more important for modern man to realize than that he is not master of his own fate.

PRE-ADVICES AND DISCUSSIONS

ASPECTS OF FREEDOM AND RESTRICTION IN THE SCIENCES

by G. E. LANGEMEIJER

It may seem curious that the task of writing a report in which science (the word throughout will be used in its widest sense) is the focus of interest, has been entrusted to a jurist. After all there is grave doubt whether the work that forms the greatest part of what we call legal science bears a purely scientific character or even only a preponderantly scientific one. Personally I am inclined to answer this question in the negative. On the other hand there may be much to be said for choosing a jurist. In the first place, even though guidance in the application of the law, which after all is the chief occupation of the majority of students of legal sciences, is performed according to a method that differs essentially from a scientific one, yet it has this in common with scientific methods that it subjects whoever applies it to a discipline which forbids him to follow his own wishes or those of others, which he would otherwise probably like to take into account. It is true that the methods of theoretical law are flexible and it is also true that the law itself may be the object of considerable differences of opinion, it yet is subject to rules which he is not at liberty to overlook, even though there is no coercion from without. These rules are laid down in the unequivocal words of the law and in the well-established habits of administration of the law and suchlike. It is true that according to the interpretation prevailing at the present day these rules do not exclude every deviation, however small under all circumstances. It is precisely this peculiarity, namely that the possibility and degree of such a deviation always falls to his responsibility, which makes the jurist become still more aware of his methods as a disciplinary system, makes him feel still more obliged to take into account the great venerability of the law, and gives him still greater reason than others to be prepared to repulse any pressure that would make him fail in this duty. And moreover, one of the main forces that might limit the freedom of science is the law, and for this reason a jurist is certainly more particularly called upon to judge.

The above remarks help to define the boundaries of my subject.

I shall turn my attention to freedom and restraint taken in their strictest sense, and not for instance in the sense of autonomy as against influences from elsewhere. I shall not discuss therefore the non-scientific influences that make themselves felt in the pursuit of science, which may be surmised with a great amount of probability but which are not consciously exercised and which are usually not consciously experienced. I am thinking of the effects of the historical situation, the social surroundings in both its wide and its narrow implications with regard to the student of science, his personal ability, experience of life and interests. These are all factors that have a particularly big chance of exercising their influence in those very fields that are most familiar to me. In so far as the scholar is aware of this influence, it is of course an important part of his responsibility to eliminate it, or, when he is obliged to choose between two possibilities, candidly to realise he is doing so. It is also necessary that he should turn his attention to distinguishing these influences. With all the good will in the world it will easily happen, however, that the influences most essential to the situation, his milieu or his personality, will remain partly hidden to him. But whether the scholar has these influences consciously under control or whether he undergoes them unconsciously, in both cases they furnish a different problem from the restrictions that might be applied from without, restrictions affecting the possibility to serve science in the way he considers the right one. It is only the latter kind of restrictions which can be combatted not only in the person of the scientific worker but also in that of the person who might influence him. And this opposition may consist in social institutions or standards which do not only concern the inner attitude of the scholar. It is this opposition which I believe to be my first concern in the subject entrusted to me. A further limitation of my report follows from the fact that only the situation in the field of the so-called arts subjects – and that only in part – is at all familiar to me. This does not mean that I consider the propositions I shall defend as being applicable in this field only. I only wish it to be the reader's key in regard to the examples and presuppositions that I had in view.

When a subject is brought up for discussion like freedom in the arts and sciences, a principle that for a long period of time has been generally accepted as above argument – at least in much the largest part of leading scientific circles – and that yet has its roots deep in man's view on life, then there is reason to inquire whether at this moment there is a special cause for it. I believe that in this case such causes are not far to seek. To my mind there are two; they are comparatively independent of each other and yet show some interconnection. The first cause is the obvious impairment of freedom in the sciences through political forces, an impairment that has become manifest in the immediate past to a degree that would have seemed unthinkable in a fairly long preceding period. Nor does it seem as if this danger has disappeared; there is still a menace, both from the right and the left. The other cause, of no less weight in the realm of the mind, is the greatly reduced prestige of science, especially as a guide for man's individual and social life. This is owing to the ever-increasing influence of the irrationalistic ways of thought of the most diverse nature – both in philosophy and in views on life that this last half century has shown. It is self-evident that the trends I mentioned in the first place are apt to take advantage of the phenomenon mentioned in the second place. And on the other hand this latter phenomenon must make it difficult to resist these trends, even for those who do not naturally feel any liking for them.

If there is good reason then for renewing the discussion of the question of the freedom of knowledge at the present day, then I yet do not think that there is also good reason to answer this question as a matter of principle thus that there is any need of giving up so much as even a particle of the claim science has to freedom, in the sense we here have in mind. Even if we admit – and I personally certainly do so – that seen the present state of philosophy, in so far as we can estimate it, there is a reason to reckon to a large extent with data far removed from science, then that does not diminish the fact that for the present the striving after increase in knowledge and insight along scientific lines yet continues to furnish a large and indispensable contribution towards these very philosophies. Nor does the fact that less importance is nowadays attached to data of a scientific nature, than was done say fifty years ago, detract from this in any way, and at any rate the pursuit of knowledge along scientific lines remains one of man's noblest means of expression. Further it is clear that science loses every meaning as soon as it can no longer work autonomously. By this last I mean that it should only be influenced by indispensable presuppositions, which the scholar freely accepts as such on the one hand, and by the inherent requirements of scientific method on the other. It follows from this that any restrictions regarding this autonomy, i.e. freedom in science, cannot be accepted if they cannot be conclusivily justified.

If I am right then such a restriction of freedom is possible in three ways, firstly through or by virtue of the legal order, secondly through economic pressure, and thirdly through moral pressure. We shall therefore have to ask ourselves whether there are cases where one of these restrictions might justifiably be brought forward and whether they could justifiably turn the scales in the face of the high value of knowledge and the freedom indispensable to it. Besides the distinctions made above I see no reason to make others according to the nature of the restriction in freedom. In practice there is no difference between compulsion to publish against one's convictions and a simple prohibition to publish one's entire conviction. Nor yet between combating the convictions themselves and a prohibition to publish them or only part of them. The seemingly slighter restriction always merges in the more serious one. Whoever will take the trouble to reread scientific publications that appeared under totalitarian régimes by scholars of whose writings one also had cognizance before and after that period, will find again and again that only the exceptionably strong personalities were able to remain entirely silent or turned their attention to non-political subjects (both of which as a matter of fact mean frustration) or were even able to write as they did before and after that period. The others one always catches in utterances, which betray that the writer does not wish to remain silent on the subjects he cares for most, but then, becoming aware that even through what he withholds he is betraying his oppositional disposition which might spell danger to him, proceeds to pay tribute, even if it is perhaps only half-heartedly, to the "ruling" views.

Firstly then we shall discuss the restrictions that the legal order might impose. This seems hardly imaginable when we think of the legal order we are familiar with. Is it possible to conceive of a statutory regulation in accordance with which it is ordered or forbidden to give expression to a given scientific opinion? As far as I know even the totalitarian forms of government knew and know nothing of such restrictions. In itself this is a fact that one might be glad to note and that is even reassuring. Apparently the fashion to belittle knowledge has not so far affected the veneration felt for it, that even Hitler shrank back from openly uttering his authoratative word against it. Meanwhile, however, the reassurance we felt lies on an ideal plane, not on a practical one. There is no one who will doubt that the organs of the totalitarian legal orders applied sanctions, some of an utterly annihilating kind, that were directed against the scientific activities of its victims. To deny them the term "legal order" has to my mind little point, seeing the function they fulfilled. In such a case it is the legal order that can be said to restrict scientific freedom as soon as its regulations make it possible that activities in purely scientific fields are affected by methods of compulsion. This may occur by giving the executive authorities the power to take action against scientific activities of a given trend that they consider undesirable. It may also take the form of quite general powers being given for the restriction of freedom, for which no special reasons need be specified, for instance like the placing of people in concentration camps under the fascist regimes.

Science therefore can certainly not be said to be out of reach of coercion on the part of the legal order – coercion that is conceivable in many other shapes than those I mentioned as examples. The question therefore: may such compulsion sometimes be justified for reasons of higher value? This question cannot be answered in the negative *as a matter of principle*. This follows from the fact that there are undoubtedly higher and profounder values than science, like what is morally right, the indispensable external requirements for an existence worthy of human beings, and for many people also the divine destination of man. The questions then that really matter are "only" the following: can these values ever be menaced by *science?* And if so, can the legal

5

order avert this menace, and, in so far as it can do so, can it do so without on the other hand doing greater evil?

Before seeking answers to the chief questions of this report it will be a good thing I believe, to set aside one part of the question, a part that would complicate matters unnecessarily on the one hand and that is of no practical importance on the other. Let me explain. It is conceivable that science may do damage to higher values, directly because its pronouncements might make those values crumble and indirectly because some result of science, in itself neutral, may well be used for purposes of negative value. This last will make everyone think of weapons of destruction, although one might also think of examples which in proportion are microscopically small, like processes for faking documents or adulterating trade commodities easily.

I believe that it is clear that the solution in the latter case can never be compulsive restraint of science through the legal order. When scientific discoveries are applied for use in horrors of war, then it is not science, which makes this indirectly possible, but the legal order, which does so directly, which is responsible. And when criminals make use of scientific discoveries, then the legal order should turn against the criminals and not against science, unless of course representatives of the latter have themselves become accomplices and have left the precincts of science for the field of crime.

The real question therefore is whether the legal order can ever be called upon to take action against the finds of science in the case of their seeming to be calculated to undermine moral convictions, or actual love of one's neighbour, or perhaps even religious faith. The answer to this may be given as fundamentally as possible and if considered strictly as a matter of principle it is even conclusive, but seen in the light of cultural reality it disposes of the question too easily.

The answer would be that no results in science are conceivable which would have this harmful effect, because a real conflict between knowledge which is aware of the limitations inherent in its tie with the human mind on the one hand, and real morality, real charity and real faith on the other is unthinkable because they are completely independent of each other.

This answer is too simple as regards the relationship between science and religion for one thing (not to mention those forms of morality which are subservient to the latter). For science and religion may both conceivably advance opinions as being true that are irreconcilable. It is true that there are indications that both, within the field of present-day Western culture, have come to avoid this thanks to a profounder reflection about their own deepest nature and limits as well as about that of the other. However, these limits have not been fixed once and for all as regards science, and a repetition of such a conflict as in the case of Galileo, which is perhaps the most striking example, is therefore not inconceivable. To discuss who should win in such a case is futile. At the moment when the conflict seems unsolvable some people would like the decisive word to go to science, others to religion, one and all in the unshakable conviction – in the full sense of that word – of their own opinion. History teaches that in the end, when both parties start bestowing more serious thought on their own doctrines, they come to the conclusion that it was only a seeming conflict, because at bottom those doctrines did not touch anywhere. Before things reach this stage, however, many generations may have passed and a legal order which might have sided with religion and thus might have retarded the elimination of the contrast (or on the contrary, perhaps have accelerated it) is imaginable. To prove as a matter of principle that it is in the wrong, is out of the question. For this would only be possible with an appeal to the possible relativity of man's opinion at a given moment about the implications and consequences of religious dogmas, which for that very reason are unacceptable to it. Science can only make an appeal to the wisdom of the representatives of religion and the law, referring them to the fact that experience has always proved that in the end such conflicts were never more than seeming conflicts. For the rest science will have to console itself that ultimately it has always been able to carry the day, while religion in its turn can feel reconciled when it considers that it has never yet meant having to give up anything really essential. Over against this we must remember that while awaiting the outcome both religion and science may do each other great damage, and that the law, if it has taken part in the conflict at all, cannot remain unscathed either.

Does the proposition that in the case of full realisation of its own nature and limitations conflict is not possible altogether hold good for the relationship between science as against public morals and practical charity? For instance is it inconceivable that science, though an overwhelming amount of historical ethnological and sociological material should "prove" that all morality is only an epiphenomenon of purely factual relationships in society, and that whoever should manifest greater love of his neighbour than would be the result of coercion on the part of his social surroundings, would thus be made to see the vanity of his ideals? To my mind that is indeed inconceivable. Even if only for the fact that if someone actually exacts higher moral demands of himself than his situation in society exacts of him this would refute the whole evidence. So either the evidence fails in this way, and therefore in other words it is not science that has come into conflict with morality, or otherwise the morality that seems to be menaced is in reality in all its various manifestations, including those whose causal *determination* cannot for the time being be traced, a causally determined product of circumstances, and is therefore in no danger, even when those circumstances are laid bare.

Thus far it was a question of a clash between real science and real morality! The development of the problem, however, already points in the direction of the conflict that *is* conceivable and that may even be of great practical importance, namely the conflict between science and morality neither of which are genuine, but which cannot immediately be unmasked for what they are. It is needless to say that science may be in error but morality may be too. I mean it in this sense: someone believes he is accepting moral standards for no other reason than that their contents appeal to him (whereby any refutation by the facts would be precluded). In reality, however, he would be driven or at least sustained by the idea for instance of social evolution, which would ultimately have to lead to the realization of, or an ever nearer approach to, his ideals; or perhaps again by an ideal order of things which is more real than so-called reality. If the foundation of such a morality, of which the person in question is unaware that it has a foundation of a non-moral or only partly moral character, is shaken it is conceivable that the moral convictions will also be disturbed.

If it is *real science* that shakes some morality that is not entirely genuine (I mean if it draws part of its strength from ulterior thoughts that are not purely moral, as its formulation suggests) then I believe it is futile to combat it. If morality is more than simply a demonstrable and necessary by-product of social reality, then it will also appear that when the props of an evolutionary or idealistic and metaphysical nature (at least those of a not strictly moral nature) fall away, a temporary disturbance may be brought about, but a disturbance that must be followed by a recovery of equilibrium.

The only case that is of truly current interest is the menace of pseudo *science* to genuine morality. Is there even then reason for the law to interfere? Might the legal order of the Weimar Republic, to mention a concrete example, have contested the racial creed in so far as it made its appearance in scientific guise?

I believe that the answer must be: No! Even if only for purely opportunist reasons, namely that it is almost certainly unnecessary or useless, while the disadvantages would predominate in the small margin of cases where it might succeed.

Legal opposition would be superfluous if the moral powers of resistance in the people were sufficient. This is self-evident. If the moral powers of resistance are not sufficient, however, for opposing what is essentially immoral, then that can only mean that the nation is swayed by passions that would never be curbed by any legal order, at least if that order wished to remain true to its own standards.

It is in the nature of things that it is conceivable that the moral and irrational forces in man would just about keep each other in balance. In that case indeed it depends on the determination of the representatives of the legal order to let the former remain in the ascendant. Also in this case however, it would be very difficult to imagine that the movement one should wish to suppress – a thing that in principle would be quite right – would draw an appreciable portion of its strength from the quasi-scientific part of the propaganda being carried out for it. Where evil passions and moral sense come into conflict with one another, it is a duel of giants. And it then makes little difference whether one of them has put on the cardboard helmet of a quasi-scientific theory. One might think, however, that it is yet always worth while to try and knock off that helmet, which apparently means something to him, even if only to shake his morale. That would be true if the imagery I have used – the triviality of which I hope the reader will forgive me – were not much too simple in relation to reality. Of course, if the legal order embodied pure morality, than one would not wish to deny it the right of suppressing pseudo-science that places itself at the service of evil passions. No system of law, however, represents morality pure and simple. It may seem to be morality of a high order in comparison to that which it is combating, nevertheless it will be represented by fallible people with all the usual prejudices and selfinterest. When once its representatives assume the right to judge whether science is genuine or not and whether it is undermining public morals, then you may be sure its condemnation will not be confined to pseudo-science, which *is* indeed immoral, but that it will all too soon also attack real science which only *seems* immoral, or which simply does not happen to fit in with their own interests. This is a disadvantage that will be much greater than the always infinitesimal advantage there might be in combating certain forms of pseudoscience, which aids to movements, which as a whole are a great danger to public morals.

So as to the rôle the law is to play, one can only be led I think to the following conclusion: it is the privilege and duty of science to protest against every form of restraint put upon its freedom from that quarter, even if the restrictions in question were prompted by the idea of serving still higher values than those of science. Science can bring forward reasonable grounds to support its protest, except if the value the law is championing, happens to be a dogmatic religious conviction.

The problem of economic pressure is quite different. That economic pressure must be rejected, follows from the fact that it is the result of forces which are not subject to any control of a higher principle. The difficulty here lies in recognizing the pressure. It is clear that here we have to deal with phenomena of a very varied nature. They extend from the absolutely objectionable cases that merge without any very sharp demarcation line into the unobjectionable decision as to how one shall make the most of one's private expenditure. The two extremes here might be for instance that a private owner threatens to sack an employee, and even reinforces his threat with the announcement that he will see to it that his employee will not find work anywhere else either, because the scientific publications the latter accomplished in his spare time were not to his liking, and, on the other hand, such a thing as that, apart from any special activities against it, a historical book, in which the writer places a national hero on a somewhat lower plane than is usually assigned him, finds few buyers.

There is one dividing line which may be drawn through the field we have here discussed, namely the line that runs between acts that take place with a more or less conscious, distinct or implicit intention of interfering in the autonomy of science, and those that may only show as pressure but where the man acting does not think of that effect. The former deserves to be disqualified from a moral viewpoint, both because it shows no respect for the high value of science, as well as because it tries to make others act from some despicable motive. The latter on the other hand does not deserve such disqualification, unless perhaps for a subsidiary reason, for instance for the narrow-minded chauvinism that expresses itself in a refusal to buy a given book as in the example mentioned above.

To my mind only the economic pressure that is meant as such belongs to the subject that is here under discussion. It is true that there may also be reason enough to fight against the sort of economic pressure that is not meant as such. But, in the efforts of trying to bring the public, or a certain portion of the public, to gain a better insight on this point, it is not justified, nor is it proper to start blaming the people in question for disregarding the freedom of science.

The question about the pressure we have in mind, is what can be done against it. It is naturally desirable to a certain extent to attest to the nobility of science at times when an anti-rationalist trend in the culture of the day, or the unavoidable waning of the value put upon spiritual goods, that have been long and tranquilly enjoyed, tend to diminish every feeling for science. Meanwhile the sort of people who are able deliberately to force a scholar into renouncing his scientific calling or conviction, will only be amenable to such arguments by way of exception. In striking cases public opinion may perhaps be aroused and then be able to emphasize such arguments which they do not possess of themselves. This too, however, will usually be an exception and more-over has the disadvantage, which should not be under-estimated, that science may thereby be involved in controversies that will not always be conducted in a style worthy of its high calling. There is reason to ask oneself therefore whether in this instance it is not the task of the public authorities to protect science against impairment from without, by the side of their duty to respect the freedom of science themselves. It is true that it is not easy to imagine how this would be possible if it had to be done as a direct combating of the pressure that is being brought to bear. Indirectly, however, the authorities nevertheless have the possibility of ensuring, or at least furthering the immunity of the scientific worker from economic pressure. This may be achieved by seeing to it that every serious student of science is – if he desires so – assured of an employment in the public service, in-volving a decent standard of living in proper proportion to the economic conditions generally prevailing in his country and time.

I am aware that the calling intended by me for the authorities has very far-reaching consequences. For does this not involve that in establishing scientific public offices (or other forms of scientific work that are remunerated by the public authorities, i.e. government grants for study and suchlike) it will not always be possible to confine things strictly to what is indispensable for making the scientific public offices function properly. Up to a certain point authorities have been acting more in this spirit here in Holland since the war. There are perhaps a few, even if only a very few, examples of professorships the institution of which was largely owing to a scholar being available, who was except-ionally fitted to occupy such a chair, and who was not appointed because of the immediate necessity the teaching of the subject in question had in the whole of some university course. For functions such as lecturer and reader and the like even more is done in this direction as far as I can see, even if there too, only with the greatest caution. And finally the subsidies of Z.W.O. (Zuiver Wetenschappelijk Onderzoek, Pure Scientific Research) and T.N.O. (Toegepast Natuurkundig Onderzoek, Applied Physics) form a further addition of this system but only of temporary nature.

Meanwhile all this is by no means a guarantee that the authorities will ensure a suitable minimum standard of living for all scientific workers. And it is natural that people should object to the idea of such a guarantee, even if only by way of a declaration of principles that give no rights to the individual. Would not such a line of conduct – it may well be asked – tempt a larger part of the population than is justified towards scientific work? I do not believe that we need be unduly afraid that this will happen. The number of those who are able to accomplish original scientific work of real value, will probably always naturally be fairly narrowly circumscribed. It is certainly conceivable that support on a more lavish scale to those of talent among the new generation of the poorer classes would raise the number. We should even hope so. But if we maintain the rule that what matters is whether the scientific worker is going to practice his subject seriously, then the danger that people might seek a sort of state pension cum light and pleasant occupation in this way is purely imaginary.

I spoke of a decent standard of living for the scientific worker. Perhaps it is not superfluous to point out that for this purpose his leisure time should be such, that he can follow his calling in so far as his work does not coincide with his scientific calling.

Thus far I have spoken first about the freedom that is due to science, on the part of the authorities, and next about the attack on that freedom by individuals possessing economic power, against whom the authorities should protect science as much as they can. There is a transitional state, however, that we have now and again touched upon in the foregoing, namely where the authorities act in the capacity of "employer" to the man of science. In many countries, Holland included, it will even be more important than that of the scientist being economically dependent on private support. In those countries namely it is the exception for universities and other institutions that serve science, to be the result of private initiative. (And then only if they have been founded in connection with some denominational standpoint) or been endowed in large part through private grants.

It is now clear that public bodies are in a position to exert the intentional and the unintentional sort of economic pressure. For instance through their decision to appoint someone or not in a given case, or by granting or withholding grants, the great difference with the way private individuals use economic influence being, that then fairly efficient measures are possible, also against the unintentional influences. The preventative is a sufficient and evenly divided influence on the part of the representatives of the arts and sciences themselves where appointments and subsidies are concerned.

A much more difficult problem lies in the possibility of some government, which avoids exerting direct pressure by means of the legal order for and against certain scientific opinions, yet makes its influence felt in the way in which it manages its right of appointment and subsidy. In principle we can soon be agreed on this point. Pressure of this sort is hardly less palpable than real compulsion, and therefore also hardly less injurious. Moreover, because of its more or less veiled character it is of an insincerity that is unworthy of public authorities. Yet it is impossible to escape from having to acknowledge restrictions with regard to this principle. When once a government considers it its duty to keep certain things secret for the sake of the highest interests of the community for which it is responsible, then no one can oblige it to give access to those secrets to people whose opinions are no guarantee for secrecy. In such a case discussion as to whether secrecy is desirable is possible, and also about the aims to which this secrecy is in turn subservient. But once the decision has been made by those who have to decide on behalf of a people, then the reproach of taking the bread out of other people's mouths directed against excluding untrustworthy people from the functions meant here, is out of place. Here then is an undeniable restriction on the freedom of science. This does not alter the fact that we will have to be fully conscious that we must stand on guard by the breach that has been made through this restriction, for terrible abuse may creep in that way: it may namely start a witch hunt. The people taking part in it may only consider a person's unconditional assent to society as it guarantee for his trustworthiness, and they may not be able to realize the moral value which is not a less but probably a more unshakable guarantee in the person of critical mind.

This by no means exhausts the problem. Is the discrimination mentioned above, the only one to which the government is empowered when it comes to granting scientific functions and oportioning tasks? Or is it also the duty of the government to guard against the penetration in those fields of scientific research and university education, that can hardly be distinguished from each other, of persons who are hostile to everything that is most fundamental in the existing legal order? It is not a simple question. It is not so on its own merits and for us Dutchmen it is particularly complicated on account of the experiences we went through under the late occupation. Moreover it is a question that it is difficult to separate from the judging of scientific competency, a thing that in part must always belong to the province of the government.

First and foremost then this question: may students be exposed to influences that undermine the convictions that are held by the great majority of a people, even if they may differ on a great many other subjects? If put in this way the question I am firmly convinced should be answered in the affirmative. Certainly, if some scholar of impressive stature holds views that are banned in a given community, then a strong spell may proceed from these ideas, especially in the case of young people. Nor would I like to rely on the knock-down argument that intellectuals still in training should be proof against such influence. That much the history of the last twenty years has surely taught us that it is given to practically nobody to make a reasonable and wholly justified choice of a party in the confusion of the profound, so-called political contrasts, and that no one therefore may consider himself above suggestion and propaganda. But, so I should like to ask, from what does the more dangerous suggestion proceed for the existing order of things: from the fact that the exceptional value of an opponent in a given field is recognized, even if there is some attendant risk, or from the fact that the existing order of things only expects its own preservation from an impenetrable barrier between its opponents, who differ on principle, and the rest of the population? There is more. When once a government starts making discriminations against a certain group, then it cannot do otherwise than believe that the members of that group (whom it will not assume to be exeptionally scrupulous) will mask their real minds. This assumption will bring the government to shut out not only its real enemies, but also anyone whose absolute trustworthiness does not seem guaranteed. Many will thus be shut out who do not constitute the least danger of treason, and who on the contrary, as being people of broad views, can only be dispensed with on pain of great damage to the development of science. But, you will say, what about those experiences during the occupation? What a lot of trouble the professors and students had with the few members of the academic senates who made common cause with the enemy. How much easier and how much less dangerous would it not have been sometimes to make a firm stand, if it had not been for the presence of a few professors in the faculty and senate meetings who were willing to pass on to the enemy the things that had been discussed there. That is true. But is it reasonable to think that the occupying power, if it had not found these tools, would not have seen to it that through special appointments such helpers were placed in the most vulnerable posts? And further, those who made common cause with the enemy during the late occupation were by no means all of them people of whom it might have been foreseen. From this point of view therefore witch-hunting among candidates for academic appointments is not efficacious either.

Nevertheless all this does not exhaust the whole problem even yet. There are cases, and especially so in the field of "sciences morales et politiques", where the boundary line between political trustworthiness and scientific competency is very difficult to draw. When the government in such an instance appoints some-one to a professorship, it thereby at the same time shows that it considers the man's convictions, the "political" ones included, to be above reproach. A government which acknowledged a national socialist before the war or a communist now as the most suitable man for a professorship in ethics, jurisprudence, polilitical science, social philosophy or even international law, would almost certainly make people doubt the spirit in which such a government that had done so or was about to do so, upheld or upholds democracy in the sense that we attach to it in the West. To require a government to close its eyes to the creed of the person who is to fill such a chair when that creed is utterly hostile to the social order upheld by that government, is to require unprincipled conduct on its part. The consequence is of course that one should not reproach national socialist or communist governments with intolerance, because they do not fill such chairs with adherents of democratic ideas in the Western sense. Yet the objections against censorship on the part of the government in academic matters have not been eliminated by this consideration. One can therefore not vigorously enough assert that the government should confine itself to what is strictly necessary, both as to defining the number of posts in question, as well as to defining exactly what ideas would debar a person from such a post.

And finally we come to moral pressure. But little need be said about it. Moral pressure will not easily occur where all legal and economic pressure are lacking. If the moral disqualification of certain aspects of science are of an appreciable intensity (I am only thinking here of real science, for the "sham" sort I refer the reader to pages 7 and 8) then economic pressure will not be far to seek and legal compulsion will probably be at least openly canvassed. Meanwhile exceptional situations are conceivable- here again a historian who should offend national prejudice at once occurs to me as the most likely example; although ethics is of course also a field where this sort of thing is readily imaginable – in which the legal and social standards of a community imply such an unshakable respect for the freedom in science that nobody will dare think of legal or economic compulsion, but in which nevertheless, a scholar may have to endure such moral disapproval that even a man of character may feel greatly restricted in his sense of freedom by it. From the foregoing it follows that this phenomenon too would deserve unconditional condemnation. Where real, genuine science is concerned, aware of its own limitations, there can be no question of its coming into conflict with genuine morality, and therefore some unsound component part must lurk in the moral condemnation. It is also clear that legal order cannot proffer protection here – apart that is from a few cases where it might come to slander, libel or insult. In this instance therefore the solution of the problem lies in the first place in removing the cause of the evil. In other words on the one hand making science realize that it must be on the watch against suggesting consequences that cannot be justified, and which would reach beyond the field of science; and on the other hand helping to refine the moral sense, enabling the latter to become aware of its unassailability by science. And for the rest to science itself falls the task of pointing the way. When once it is accepted as part of the scientific spirit not to fight one's opponent by coupling moral disqualifications to his scientific opinions, then the public at large will not be so soon inclined to do so either. Here again it is the "sciences morales et politiques", those sciences in fact that cannot do without evaluating views, that will find this self-discipline the most difficult. I could give examples to show how it sometimes fails, even with personalities of the highest order.

To end with I should like to make some brief remarks about a domain that lies on the dubious boundaries of this subject. In our time equal chances of serious views finding an audience depend on the disposal over the usual means of publicity, all more or less controlled by the government, the radio, and soon also perhaps television. In most cases sciences will not aim at spreading its views among the largest possible audience. It may happen. however, that science does think it of paramount importance to reach the largest possible public because there may be results that it considers should immediately influence the behaviour of the public in some specified way. Where this is the case the fulfillment of its vocation brings along with it that it should be given the opportunity of doing so. For a legal order that wishes to accord full rights to the freedom of science it will therefore also be part of its duty to see to it that the practitioners of science who wish it should be given the opportunity to make use of the wireless, with in principle no more control from the government than alone on the scientific merits of that which they wish to broadcast under the auspices of some scientific body.

The argument I have ventured to lay before you is not what might be called "high-sounding", a thing one might have expected at a congress, which does not so much have in view that we should try and convince each other of things on which we probably do not differ very much anyhow, but rather that we should come and testify to our scientific calling and the freedom we must claim for it in order to be able to realize its deepest significance. However, such testimony will be sufficiently perceptible in the arguments of others who certainly need no additions from me. That is why I thought I might give what as a predominantly practising jurist I could perhaps best give, namely an examination in as matter of fact a way as possible of how and how far freedom in science may be practically realized and ensured.

Nevertheless there is a conviction of principle underlying my argument, namely that principles in human society, like that of freedom in science, cannot lightly, if indeed at all, be called absolute. But the relative value that may be attributed to such principles within a measurable time and for measurable circumstances after serious thought, that value may be high enough to enable us to make a firm stand against attack from without and against any trifling doubts that might assail us from within.

DEBATE ABOUT PROF. G. E. LANGEMEIJER'S PRE-ADVICE ON "ASPECTS OF FREEDOM AND RESTRICTION IN THE SCIENCES"

by I. SAMKALDEN

Our colleague Langemeijer reveals at the end of his report the fundamental conviction which underlies his reasoning. The elements of human society, such as freedom of science, though not absolute, have a relative value, high enough to make one take a firm stand against rash interference from outside this society and against inconsiderate doubt from within. Though I have less objection against calling those elements of society of absolute value, Professor Langemeijer's conviction is also mine. Perhaps this gradation is the reason why I am of another opinion as Professor Langemeijer on one, though not an unimportant item.

On page 7 Professor Langemeijer discusses the question whether the legal order may interfere, when pseudo-science threatens real morality. His answer is negative and this "no" has a challenging touch, linked as it is with the very concrete question – given as an example – whether the legal order of the Weimar Republic should have had a right to fight the race doctrine appearing in scientific disguise. I accept this challenge. My answer is yes, the legal order certainly has a reason for interference in such a case.

President Thomas Jefferson said in his maiden speech the words which nowadays would sound to us like a fairy-tale: "If there be any among us who would wish to dissolve this Union or to change its republican form, let them stand undisturbed as monuments to the safety with which error of opinion may be tolerated where reason is left free to combat it". To say such a thing, one cannot but have an invincible confidence that by reason – and only by reason – one can find the truth and that the truth convinces. Happy the period and happy the man, who can have so much confidence; we cannot have it; at least not in such an absolute way.

Scientific work does not any longer mean finding the truth and preaching the truth and nothing else. We must admit, that it can hardly be more than a serious attempt to approach the truth as far as truth may be understood rationally. In this uninterrupted striving at the truth we are continuously aware that the result may be wrong; indeed, without this relativity scientific work is hardly imaginable. The conviction that one's own point of view is the right one, never makes one deny the possibility that somebody else may be right also. This relativity causes another one. Not only truth is convincing; so may be untruth if only disguised as truth and vested with its authority. The corruption caused bij pseudo-science may be large, even irreparable, much larger than the image of the cardboard helmet, used by Professor Langemeijer in this connection, would make us believe.

The disguise is surely not meant for a fight but merely to force one's way into the garden of science. Once there, the damage may be immense for in our garden everything is intended to be open for reception and nothing to be closed for defence.

Why should a legal system take action against a man who, using public authority as a disguise, became known as Captain of Köpenick and why should it take no action against someone who causes larger moral damage by abusing science's authority? There are examples of concrete situations in which people, whose regard for their fellowmen and for science is above doubt, acted otherwise than the preadviser recommends in the abstract. A committee of the Royal Netherlands Academy of Science recently asked for legal measures against pseudo-science in the field of cosmic rays. At Leiden, twenty years ago in 1934, an International Students Congress was ended prematurely by the Vice – Chancellor because this representative of legal order in the university community refused the German national-socialist Von Leers admittance to the university. Vice-Chancellor Huizinga accounted for this interference in one simple sentence in his valedictory address in September 1934: "A university sometimes finds itself obliged to take action simply to defend the sacred intellectual grounds on which it stands". As far as I remember, this sentence is the only one, which during the relative monotony of this annual report, ever made a packed auditorium rise in an almost frantic applause.

But though in my opinion the legal order should be used against the menace which pseudo-science may mean to fundamental moral values, I fully acknowledge the serious objections from motives of expediency, which Professor Langemeyer connects with such action by legal authority. The Captain of Köpenick can be exposed because the authenticity of his qualifications can be checked with known data. But is there such a characteristic to mark the dividing line between the serious approach to truth on one side and untruth disguised as truth on the other side? Yes, Mr Chairman, in my opinion such characteristics exist, and it is mainly due to this conviction, that Professor Langemeyer's objections of expediency do not settle the question for me.

The characteristic of pseudo-science, which is so dangerous for the moral values, is its pretended absoluteness, its intolerance towards other opinions, the evil with which it threatens its opponents. Wherever these phenomena appear, the danger signal has been given. And wherever the danger signal has been given a legal order, which has to be the guardian of our highest moral values, should be ready for action.

PROF. LANGEMEIJER'S ANSWER
TO PROF. SAMKALDEN

Colleague Samkalden has rightly said that the passage against which he raised objections was challenging. It was already so on account if my having purposely chosen the most striking example.

I wish expressly to state at the outset that there is no connection between that passage and the certain amount of relativism that is apparent in the last paragraph of my pre-advice. This relativism does not exclude an unconditional choice of party in a concrete situation, and my rejection of the racial doctrine is equally unconditional as that of colleague Samkalden.

I do not think, however, that the opportunist considerations, on account of which I assign no task to the legal order here, have been refuted by colleague Samkalden's examples. He already admits this himself in the case of the captain of Köpenick. The Von Leers case is indeed illustrative. However, this makes me wish to ask on the one hand whether, (even though vice-chancellor Huizinga acted formally as a man who at that moment happened to have a say as to who should enter the University Building or not) whether the excellent effect and the authority of his deed did not rest wholly on the fact that he was Huizinga, and especially that he was aware that he spoke in the name of the whole Dutch academic world? And on the other hand I should like to ask whether one should really believe that the organs of the legal order, which do not necessarily offer the same gurantees in their set-up, will be able to handle the criteria of pseudo-science, which colleague Samkalden mentions and which in themselves seem accurate to me, without the dangers being greater than the advantages?

DEBATE ABOUT PROF. G. E. LANGEMEIJER'S PRE-ADVICE ON "ASPECTS OF FREEDOM AND RESTRICTION IN THE SCIENCES"

by P. J. BOUMAN

The debater who wishes to contribute in bringing a pre-advice up for discussion, will sometimes be in doubt as to whether he shall direct his critical remarks against a few passages in the argument, or whether he shall point out some gap or other. The one as much as the other may easily give rise to mere carping.

For me the choice is not difficult. I can agree entirely with Professor Langemeijer's train of thought. I admire his lucid exposition and accept all his conclusions. Thus it only remains to mention what in my opinion is a lacuna – and I should like to remark that it has relation to a point of general significance that is of importance to the whole subject we are here considering. It is a case of his being inclined to delimit the concept of freedom defensively against every machination against it from outside. Some of the pre-advicers have included problems of a psychological nature in their survey, and especially the responsibility of the researcher, but I refer to something else.

Professor Langemeijer gives a restriction, which in itself is warranted enough: "I shall not discuss therefore the non-scientific influences that make themselves felt in the pursuit of science, which may be surmised with a great amount of probability but which are not consiously exercised and which are usually not consciously experienced. I am thinking of the effects of the historical situation, the social surroundings in both its wide and its narrow implications with regard to the student of science, his personal ability, experience of life and interests".

This restriction seems right and necessary to me, but when I then see that all his attention is directed towards threats against freedom that come from the outside (I, through or by reason of the legal order, 2, through economic pressure, 3, through moral pressure), then I cannot help but wonder whether the restriction in question does not throw away the baby with the soapsuds. In between the "non-scientific influences" and the pressure from without lie a certain amount of problems which are not touched by the restriction that was made and which are not covered by the questions that were treated.

I believe there are *internal* problems that are inherent to the pursuit of modern science, which cannot be ignored when it comes to a survey of freedom in science. Huizinga has more than once given warning against what he called "the mechanization of civilization". With him too, however, it was especially the dangers from without that he had in mind.

Is it not possible to point out trends showing an inner inflexibility in the pursuit of science? The story about the Trojan horse describes more than just an interesting story out of the Trojan wars. It symbolizes a tragedy that characterizes all human activity, a dialectic element of progress that calls up antithetic forces which ultimately lead to frustration. Whilst we peer over the wall at the advancing hordes of those trying to ensnare science and while we make a stand against every restriction of freedom in science, we are apt to lose sight of the internal dangers that are threatening freedom just as much.

Modern science, for its proper practice, needs an ever increasing machinery. It requires an organization which may easily develop in over-organisation. If I previously spoke of Huizinga's ideas about the "mechanization of civilisation" then I am now reminded of Max Weber's magisterial reflections about the impossibility of averting modern bureaucracy (I use the word in its current sociological sense). The freedom of the scholar is beginning to get entangled in a machine of which it is ever more difficult to gain a clear view. My Groningen colleague Arens the chemist, spoke the other day in his oration of the "joy and despair" in science. Despair may well overtake anyone who tries for instance to get some idea of the amount of scientific publications. Arens states that in 1953 there appeared 62000 publications in the chemical field, whilst it was only 5000 some 50 years ago. It is of course possible to find a way out through specialisation, but by doing so one only gains a smack of freedom within a limited field. It will not always cause feelings of despair to vanish.

May I continue my remarks a bit further? – accepting the restriction made by Professor Langemeijer that we shall leave "non-scientific influences" out of consideration.

Together with the inflexibility inherent in the modern pursuit of science other phenomena than the already mentioned extension of the general machinery can be pointed to. I should especially like to call attention to our ways of organization, for instance to the bane of the present day: our prediliction for boards of all sorts. For every advice, for every preparation for a series of decisions, in fact for any and every form of activity boards are appointed, preferably many in number, on which sit many men of science. It is of course possible to refuse an appointment, but when all is said and done one should not, as a university man, refuse to take one's place in a senate committee. Nor can one withdraw from activities pertaining to committees that pave the way for subsidizing one's own scientific research work, nor from committees that are connected with the existence of scientific institutes. But this is not all. There are editorial meetings, conferences, congresses, etc. And always there is the legitimation that they are useful and important, but taken in their totality they urge the man of science on to a degree of activity that deprives him of what is ultimately indispensable for pursuing science in full freedom: peace and reflection. To say it vulgarly one might express all this in the question: "why are we all so busy?"

Less and less are we to be found in our studies, less and less are there deep-going talks between colleagues on their special subject. We sacrifice a piece of our freedom voluntarily to enter the treadmill of an ever-increasing number of duties. A shift in accentuation may be observed in our sense of responsibility: and increased sense of care for the extension of the machinery and a weakened inclination towards intensive study. Certainly, there are arguments enough in favour of all this: co-operation, planning, international contact, etc..They provide us for instance with very good motives for the increasing numbre of international congresses. But, mindful of the old Dutch proverb that one should not speak of the gallows in the house of the hanged, it is better if in these surroundings I hold my peace on the subject of congresses.

You understand my meaning. I must be brief, but I believed that by the side of the "non-scientific influences" that were eliminated by Prof. Langemeijer, I should point out a few phenomena that are dangerous to the freedom of science. If we cannot come to the point of seeing our own faults and if we only make a stand against threats against our freedom coming from without, than we shall imperceptibly sink further into the bog of over-organization. Please forgive my perhaps somewhat black-hued reflections (partly connected with the special vulnerability of the much-invited sociologist), but I believed it would be a good thing to bring forward with the necessary frankness a few points that might enliven the discussion.

I should like to put one more question: What does the pre-adviser mean exactly by the "sciences morales et politiques"? Do not all the humanistic studies fall under that category, including history and philosophy? If this is the case should not this lead to the extension of the list of professorships on page 76 whereby the political creed of whoever were appointed would, according to the pre-adviser, have to be taken into consideration too.

PROF. LANGEMEIJER'S ANSWER TO PROF. BOUMAN

I used the term "sciences morales et politiques" as the best way of indicating a group of sciences not easily described in any other manner now very clearly definable. I should like it to mean to include those subjects in wich an appraisement of human behaviour plays in important part. To define more exactly what I mean I should like to say that I would not include history (even though in France historians *are* members of the Académie,whence I borrowed the term) nor yet all the various branches of philosophy. As a matter of fact it was not necessary to give a very sharp delimitation. For my list on p. 76 is meant as a still more circumscribed circle within the one already mentioned. I should wish to make a stand against having the circle widened. The only justification of what I there argued is that the government cannot be required (when it comes to chairs for subjects where the integral element, a profession of values, preponderates) to consider someone the best candidate for the vacancy who is squarely opposed in his profession of values to the large majority of the population.

As regards the other remark made by colleague Bouman I can only say that I whole-heartedly agree. The number of committees and things of that sort that high officials, and perhaps other categories too, have to attend is indeed not much less than for professors. I certainly believe something should be done about it. However, there is some good reason for this phenomenon too, but finding some solution is definitely beyond the scope of this congress.

ASPECTS OF FREEDOM AND RESTRICTION IN SCIENCE

by W. F. WERTHEIM

When we speak of "freedom" and "restriction" in science, these terms are almost always taken in their formal sense. The term "restricted science" calls up associations of historical examples of interference in scientific investigations through the arbitrary word of powerful authority. It makes us think of personalities like Galileo and Copernicus, whose work was hindered by church dogmas, of the prohibition for teaching Darwin's theories, which prevailed not long ago in various parts of the world and which apparently still holds good in a few states in North America, and of the Third Reich where there was nor room for the theory of relativity, nor for its creator. Or, to remain nearer home, of the impediments that the spread of knowledge about possibilities for birth control has to contend with in Holland. Science is called "free" when the investigator or scholar is permitted to perform his investigations and to publish his results, and when there is no direct prohibition for applying those results. "Freedom" in science is thus understood in a negative sense. Because prohibitory measures on the part of the authorities are lacking, the investigator or scholar "may" inquire or write what he likes. Formally he is free. It is another question, however, how far that freedom really extends. And, still more important, in how far science is free to develop its full potentialities.

This formal way of thinking neglects an extremely important aspect of freedom and the lack of freedom. It is not only a question of what the man of science is allowed to do in theory, but also what he is actually able to effect. The formal conception of freedom is related to the nineteenth century idea of it, which contented itself with the legal possibility for workers to enter into employment or not as they wished, and to avail themselves of all the blessings of the industrial revolution. Just as legal freedom for the worker might be coupled with extreme economic lack of it, in the same way formal scientific freedom is in no wise a guarantee for material freedom in the sense of development to its full scope for science and for those who practise it.

When speaking of material freedom, we start from the idea that science, including its constant development, represents a positive value. This idea is the governing idea at this congress, notwithstanding the considerations devoted to the limits of science. It is futile to champion scientific freedom, if one does not value science, taken as a whole, in a positive sense. Our starting point should be that mankind owes its whole fund of culture and civilisation to the processes of technical and scientific progress. Even if one should wish to enter the negative aspects of this development on the debtor side – apart of course from the subjective character of every evaluation of progress as "positive" or "negative" [1] – then one should realise that without knowledge and science we would still be in a state of complete subjection to nature. It would not even be possible for us to take stock of such problems as those here under discussion. Perhaps man would then in a certain sense be less burdened with guilt. But his innocence would then be the innocence of ignorance – of the animal. Now that man has eaten of the tree of knowledge, however, there is no way back to this paradise state of vegetating.

But our knowledge has made such progress that we at the same time come to realise that the blessings of science and technics are divided very unequally over the various peoples, making up mankind. The more we learn of the state in which man lives, who has hardly yet tasted of the tree of knowledge, the less we long to be back in the pre-scientific, "primitive" state. And we realise the more too, that it is not sufficient to give science free play in the formal sense. We then start feeling it as our task to develop science and technology to their fullest scope. Seen in this way science can only be called truly free, when it is offered the fullest opportunity of fulfilling itself. Science must be called "restricted", when its progress and self-realisation are hampered by all sorts of factors that cannot be explained from the scientific methodology as such.

On closer reflection it appears that such frustrations [2] are anything but confined to the so-called under-developed worlds of

---

[1] Cf. the treatise on the "idea of progress" by Prof. J. Romein and Prof. F. L. Polak for the Netherlands Sociological Association, Sociol. Year Book IV (1950).

[2] See for example "The frustration of science", by Sir Dan. Hall and others, London, 1935.

Asia, Africa and Latin America. In our Western world the scientific and technical progress made since the industrial revolution was so rapid, that we are inclined to consider the rate of development as the maximum, and even to believe that the optimum has been exceeded. Many would prefer the rate to be slowed down rather than speeded up, since they fear that a too rapid development would cause a disturbance in the equilibrium, a disturbance they even believe they can already distinguish.

And yet one might maintain more rightly that even in our Western world the development is in many respects too slow rather than too fast. Disturbances in equilibrium are caused by an unequal rate of development in the various branches of science and technics. The structure of our society brings along with it a rather one-sided technical and scientific development. This one-sidedness is most clearly seen in the following figures, which have been taken from official publications in the United States.

The total sum spent there by the government, industry and universities together on research and development was quadrupled between 1941 and 1952. The Federal Government's share in these expenditures rose from 41 % in 1941 to 60% in 1952.

These figures in appearance seem to be a striking confirmation of the opinion that science is developing in all its sectors at a breath-taking rate, and with full government support. But the picture changes at once as soon as one knows that of the funds spent by the Federal Government in 1954 about 90% was expended on research and development for military purposes. 76 % of this was the share of the Department of Defense, 10 % of the Atomic Energy Commission and 4% of the National Advisory Commission for Aeronautics.

The ratio becomes clearly manifest when we know that the research for non-military purposes financed by the Federal Government was a little more than quadrupled between 1940 and 1952, whilst its expenditures for military research and development during that same period had grown by forty times.

Besides this it is significant that of the Federal expenditures about 94% went to applied research and development, and only 6% to fundamental research.

A growing part of industrial research was financed by the government. The share of the universities in the sum total of scientific research is extremely small (only 2%). Of the research funds, expended by the Federal Government in 1951 on non-profit institutions, 88% were administered by the Department of Defense and the Atomic Energy Commission. Only one fifth of these monies was spent on fundamental research [1]).

From this example it clearly appears how one-sided and unequal scientific development has become in the West. A one-sided development necessarily means that in many fields the work will be in arrears, and that scientific advance, as a result of external factors lying outside of science, falls short of what might have been attainable under more favourable circumstances. That the work in various branches is behindhand may be inferred from the disproportion in the financial backing of the research. Thus the review article I cited above points to the proportionately extremely low sums spent on the social sciences, medical science and, as we have already had cause to mention, fundamental research.

The very limited inquiry held in Holland a short time ago by the Prince Bernhard Fund, which left out of account all scientific research in behalf of industry, shows a similar picture in so far as here, too, there are indications that in the medical and social sciences the work is considerably behindhand [2]).

On the other hand such disproportionateness in the financial field is after all no more than an indication that things are behindhand. It is of more importance whether such deficiencies can actually be proved from the angle of the sciences in question. This seems paradoxical, since one can only be certain of a step forward after that step has been taken as soon as one sees that the work is in arrears, this seems to coincide with overcoming those arrears. Yet this is not so. Even apart from the numerous instances, where the work is not behindhand owing to a deficiency in research, but as a result of not applying what is already known, one can very often predict, starting from present-day knowledge, in what direction research will be fruitful. If this were not so,

---

[1]) Taken from Bernhard J. Stern "Freedom of Research in American Science", Science and Society, vol. XVIII (1954), p. 106 ff. In note 11 the author mentions a series of official publications out of which he built up his survey.

[2]) Cf. ,,De omvang en aard van het wetenschappelijke onderzoek in Nederland buiten het bedrijfsleven 1949–1950". (The scope and the nature of scientific research in the Netherlands outside businesslife 1949/1950). Communications of the Prince Bernhard Fund, Appendix C, table 8.

then the very high costs expended by business on planned research would be so much money thrown away. One need only glance through the well-known works of J. D. Bernal and Vannevar Bush [1]) in order to realise that also with our present-day knowledge it is possible to point to the gap between "what science does" and "what science could do".

It is these external frustrations of scientific development that will be examined in the first place in this preliminary report, as being one of the most important aspects of present-day restrictions in science.

Of what nature are these external factors that impede the progress of scientific research? This question was investigated for Holland by the Amsterdam Division of the Netherlands Association of Scientific Research Workers (Verbond van Wetenschappelijke Onderzoekers, V.W.O.). The examples and reflections that follow are in large part the results of this investigation, the complete findings of which may shortly be published in a separate report on behalf of the Association of Scientific Research Workers.

These external factors then may in the first place be connected with the economic structure of our society. Lincoln Work testified that "the research division of a company is called upon to project, maintain and improve the company's position in business" [2]). From this it follows, that industry will only be interested in the sort of research, that seems profitable from a business point of view. In practice the interests of the employers by no means always coincide with those of the consumers. Although it is to the good of the consumer if there is improvement in quality and durability, the interests of the employers will in many cases be better served by moderate quality, handsome get-up and by products that are rapidly worn out, with as a result an increased turnover. Scientific research that would lead to the production of almost indestructible goods, is therefore pushed far into the background. A well-known example is the correspondence that took place between A. F. Philips and Clark Minor of General

[1]) J. D. Bernal F. R. S., "The Social Function of Science. What science does, what science could do", London 1939; Vannevar Bush "Endless Horizons", Washington 1946.

[2]) Lincoln T. Work "The philosophy and economics of an industrial research program", Research Operations in Industry, ed. by David B. Hertz and Albert H. Rubinstein, New York 1953, p. 4, quoted by Stern, l.c., p. 113.

Electric in 1934. On January 30, 1934 Philips wrote that "there seems to exist in various territories a growing tendency to supply lamps for higher voltages than in the past, which therefore leads to the conclusion that in a great many cases such lamps are being underrun.

This you will agree with me, is a very dangerous practice and is having a most detrimental influence on the total turnover of the Phoebus Parties. Especially with a view to the strongly decreased prices in many countries, this may have serious consequences for Phoebus and after the very strenuous efforts we made to emerge from a period of long-life lamps, it is of the greatest importance that we do not sink back into the same mire by paying no attention to voltages and supplying lamps, that will have a very prolonged life". Minor wrote back: "I quite agree with your proposal" [1]. This therefore is a clear example of frustration in scientific progress, owing to an *economic* factor.

In a somewhat different form this factor may play a role when a given scientific and technical improvement might lead to discharges on a large scale, because the work done by hand could then be done by machine. In that case, if the researchwork is not carried on or if the invention not applied, it is owing to a social aspect as well as an economic one.

Another variant of economic frustration is that, which is caused by lack of sufficient funds for the investigation. In so far as it is industrial researchwork that is in question, this factor largely coincides with the economic factor. The small inclination on the part of the entrepreneur to favour certain types of research, may lead to insufficient financial backing. But the financial aspect should be seen independently, when one thinks of the possibility of research-work that is backed by the government or by institutions, not working for profit. As things are today, scientific research-work is very often extremely costly so that formal permission for some investigation has very little value, as long as the practical means for carrying it out are lacking. In order to be able to develop freely science does not only need a liberal atmosphere in which the pursuit has been granted to the investigator or scholar. Without very real support in the shape of ample financial means he can get no further. One sees here how much the great stress that

---

[1] James Stewart Martin, "All Honorable Men", Boston 1950, p. 147/8.

the present-day world situation lays on research for military purposes may interfere with other kinds of research. As a result of the preponderance of research-work for military purposes there is little money left for other research-work and, what is quite as important, not enough man-power. Stern, in his essay that has already been quoted a few times, mentions a sharp retrenchment in 1953 of American expenditure for combating venereal disease that has been caused by the increase in military expenditure. [1]).

Another impeding factor that often occurs is the psychological factor, which is often closely bound up with the feeling of resistance man has to all that is new. It is easy to find examples that come not only from the past, but also from the present. Conservatism, however, is not a quality that belongs to the non-scientifically trained individual only. The conservatism within science itself may also be a strongly impeding factor. I should like to quote the well-known demographer Alfred Sauvy: "Rien n'est plus difficile a réformer qu'un enseignement, nul organisme n'est plus conservateur qu'une Université" [2]). Science not only has to put up a fight against the dogmas and resistances against what is new that prevail outside the scientific world. During the history of scientific progress science has quite as often had to fight the stereotyped truths that were thoughtlessly accepted by the circle of specialists of the various branches of knowledge. Practically every scientific renovation is the result of a fierce struggle against formal knowledge.

The last impeding factor I should like to mention, without striving to be exhaustive, is related to the psychological factor: I mean impediments of an ideological nature. Many a renewal comes up against resistance that springs from this source. This holds good just as much for new scientific results. As an example I only wish to mention the ideological opposition to artificial insemination, which here and there is even extended to its application on animals. The opposition might just as well be directed against the research itself as gainst the propagation on a large scale of results that have already been discovered, and their application.

---

[1]) l. c., p. 11.
[2]) Alfred Sauvy "Théorie générale de la population", Vol. I, Economie et Population, Paris, 1952, p. 7.

All the above mentioned factors may quite as badly damage the
the progress and self-realisation of science as a formal prohibition.
In present-day society any government that withholds the
necessary means from certain forms of investigation is thereby
enabled to suppress these forms of investigation as effectively as
if it were formally to abridge the liberty of some research.

From the foregoing it has repeatedly appeared that the restrictions are sometimes connected with the scientific research as such,
and at other times with the application of the results. These two
aspects are by no means always easy to distinguish. Before a discovery can be realised along technical lines, closer investigation
and experimentation are often necessary, even if only to arrive
at a fair cost accounting. Nevertheless, the non-application of
what has already been discovered is a very striking and serious
phenomenon in present-day society. Thus Sir Henry Tizard,
at that time chairman of the "British Association for the Advancement of Science", declared at a congress of this association
in 1948: "What is of the very greatest importance, is the application of what is already known". The British scholar believed
that no single new discovery was able to have such a rapid and
favourable effect on British industry as the application of that
which was already known.

This retardment in the application of what is already known
is particularly felt to be in the nature of a *frustration*, when
patents are consciously taken out or bought up in order to hamper
application. In 1934 it appeared from an investigation by a
Federal Commission in the United States that over 2000 patents
had been bought up by the American Bell Telephone Company,
not to use them themselves, but to prevent competitors from
using them. [1]).

In cases such as these big concerns therefore have the power to
hinder the application of new discoveries. These economic, frustrating factors occur especially when certain industries or industrial concerns succeed in gaining a position of monopoly. In
the work of J. S. Martin that has already been cited a number of
striking examples are mentioned. In this way the German Degus-

---

[1]) Cf. "Technological Trends and National Policy", National Resources Committee
Report, Washington 1937, p. 50. See also S. Lilley "Science and Progress", London
1944, p. 24.

sa-concern paid some British chemical industry a yearly sum of
$ 25.000 during the years 1931–1939, in order to stop the produc-
tion of cyanides. The same concern paid a professor in Berlin the
yearly sum of $ 10.000 in order not to carry on his new method
for producing cyanides [1]). The publication of patents, moreover,
sometimes proves by no means to make the process clear which
should enable the technician to follow it up. When they tried to
make Salversan in the United States according to the patents
during the first world war, the substance they had obtained proved
to be badly poisonous. The patent dit not indicate how the arse-
nic was to be removed from the medicine [2]).

These impediments on the application of what is already known
may very well be experienced, also subjectively from the point of
view of the investigator, as restrictions on his freedom. Patents
were originally meant to stimulate the scientific research-worker
to achieve his utmost by guaranteeing him the fruits of his discovery for a definite period. But now that patents on the contrary are often used at the present day by the companies in whose
service the inventor is, to stop their application, the stimulus
easily becomes a drag [3]). Psychologically speaking too, the
research-worker will perform his work with greatly diminished
eagerness when there is no hope of ever seeing his invention used,
or if he must be afraid that his invention, socially speaking, will
be used in the opposite way from that which he in his idealism had
imagined. "There is a feeling of frustration amongst many of the
Government's research-workers", in the words of Mr Stanley
Maine, "because time after time they undertake and complete
researchwork, recognized as valuable work, frequently the work
is published to the world but nothing more happens, even when
the most o b v i o u s user of the research is another Government
Department" [4]).

But when a patent has been taken out, then at least the foun-

---

[1]) J. S. Martin, 1. c., p. 115.

[2]) J. S. Martin, l.c., p. 62/3.

[3]) See W. F. Wertheim "Stimuleren of remmen octrooien onze vooruitgang" (Do
patents stimulate or hinder our progress), Ned. Jur. blad (Neth. Law Journal), 1947,
p. 81 ff.

[4]) Cf. Questionnaire: Delays in the application of science in industry. A summary
of replies sent by scientific and technological organisations to the Parliamentary and
Scientific Committee to a questionnaire, issued prior to the debate on Science in the
House of Lords, June 11th and 12th, 1952.

dation has been laid for some future step forward in the scientific field, even though there may be a deep gulf between theory and practice, and between discovery and its application. Scientific progress, however, is arrested still more when the results of the research-work are kept secret. This occurs in increasing measure in the present-day state of world affairs. The research-work for military purposes is first and foremost enveloped on all sides by secrecy; the so-called atomic secret was only one example out of many. But the industries working for profit also envelop their scientific research with veil upon veil of secrecy. There are big industries which systematically fail to publish the findings of the research-work performed within their walls. But those industries too, which carry on an active patent-policy, reserve to themselves the right to keep strict supervision on the publication of the research-work carried out in their laboratories. Stern states that the findings published by concerns possessing industrial laboratories are meagre compared to those published by academic and government laboratories [1]). This secrecy is a bad check on the free exchange and circulation of scientific results.

But apart from secrecy there are numerous factors that hinder the effective circulation of the knowledge obtained through scientific research. The investigation instituted by the Amsterdam division of the V e r b o n d   v a n   W e t e n s c h a p p e l ij k e O n d e r z o e k e r s (Association of Scientific Research-Workers) to go into the factors that in Holland hinder the application of results already generally accepted, went to show how strong is the influence of the ignorance of the public in respect to results already discovered. Lack of the right sort of enlightenment of the public on the part of the scientific world plays a large role in all this. But the element of frustration occurs more specifically when this ignorance of the public is exploited and increased by misleading advertising that only has commercial ends in view. Since a short time the "Consumentengids", the organ of the Union of Consumers in Holland, regularly published cases of this sort of misleading advertising. As an example I shall only mention the advertising done for the Denicotea-filters, which profess to take away the damaging effect of cigarette-smoke. Research brought to light that these filters hardly had the advertised effect at all, as

---

[1]) Stern, 1. c., p. 114.

they only took away a negligible part of the nicotine out of the cigarette-smoke [1]). In the pharmaceutical world the unbridled advertising of new medicines that goes on, has a somewhat different effect. It has become so difficult for doctors to gain a clear view of the situation, owing to the enormous amounts of new medicines that are brought on the market, that the result is that the application of what is really important is considerably retarded.

It is clear from the foregoing, how important the checks on the free development of our knowledge are, which flow from the fact that what is scientifically established, is yet by no means always applied. A new complication arises in this whole problem. How is one to determine whether being behindhand in some application is j u s t i f i e d or not? And the same question might be put in respect to the arrear in the research-work. In other words: by stamping it as a deficiency when a result fails to come or is not applied, an appraisal is implied. It presupposes that one is of opinion, on the grounds of a certain vision on human welfare, that the omission of scientific and technological renewal is to be condemned from a social point of view. We are thus meeting with problems in the field of semantics. Pointing out a "frustration" presupposes that without that "frustration" the progress of science would be "possible" and in this concept "possible" many presuppositions lie concealed that are bound up with opinions of an ideological or social order.

The fact that some actual industry does not introduce a new technical process is a strong indication that as far as the interests of the industry are concerned the new process is no improvement. The question now is only in how far the critic accepts the industry's standpoint as decisive. Whoever is unwilling to accept that the general interests of the public and the interests of business are identical, and who is critical of the existing economic structure will much sooner be inclined to speak of a frustration of an economic nature, than he for whom it is enough that the new process does not pay. One might even go so far as to say that the economic structure as such is responsible for many of the deficiencies, if one thinks on the one hand of the shortage of scientific

---

[1]) Cf. "Consumentengids" (Guide for the Consumer), no 1 (April 1953), and 2nd vol. no 6 (June 1954).

equipment in the smaller industries, and on the other of the frequent contrast between the direction in which the research-work
and the technical development is conducted in the big industries
that enjoy monopolies, and the interests of the public at large.
But then again one will have to admit that a renovation that can
only be introduced by sacrificing a considerable amount of capi-
tal and installations that had been invested in the old process,
should be carefully weighed against the loss through writing off,
and this not only from the point of view of the owner who works
for profit, but also from the point of view of the public interest.
In between the obvious cases of frustration on the one hand and
renovation that does not pay on the other, there are very many
gradations, which will be judged differently according to the
point of view. In the same way the man who considers the military expenditures a necessary evil, will not so soon be inclined
to look upon the financial factor as a frustration, as the man who
considers the armaments-race as senseless.

In the foregoing, attention has chiefly been paid to cases of
frustration of so extreme a nature, that there could hardly be any
doubt among men of science as to the correctness of the label. But
if one is agreed about essentials, then it still often proves extremely difficult in practice to point out unambiguous, concrete cases
of frustration, because the reason why a certain scientific renovation was not applied, is usually bound up in a tangled knot of
prices and business-regulations, that is most difficult to unravel.
One of the forms in which frustration often occurs in practice, is
the exorbitantly high price, that is asked for the product made
according to a new process, which leaves the improvement in
question out of reach of the general public. But how can one prove
on the one hand that the high price limits the sale, and on the
other hand show with cogent arguments that the price of the
product – for instance an antibioticum – has been unnecessarily
forced up? Added to this it is very difficult for outsiders to find
their way in the maze of commercial regulations and considerations. Investigation into this matter is hindered through the
secrecy we have already mentioned before, which envelops laboratory work in industry. The real reason why a certain decision is
made at the top, often remains hidden for the scientific workers
as well. And if this is not the case, these people will be on their

guard against openly speaking about such frustrations with outsiders, since by doing so they might seriously endanger their position. This "silk curtain" that envelops a large part of laboratory work might in itself be called one of the chief frustrating factors in our society, because on account of it one can hardly make out how scientific progress really stands, and what chances it has missed or is still missing.

Only if inquiries are instituted by the government, supported by sufficient means of constraint to furnish the investigators with opportunity of penetrating into factory secrets, only then will there be a chance of the curtain lifting. Our most important data about frustration in the Western world we still draw from the inquiry in the United States by the National Resources Committee that was instituted by order of the Federal Government in the thirties, and which led to the famous report "Technological Trends and National Policy" (Washington, 1937). An inquiry like the one held by the Verbond van Wetenschappelijke Onderzoekers can at most only lift a corner of the curtain. Moreover part of the established data that have been discovered cannot be published out of considerations of safety for the research-workers who furnished the information. Nevertheless, notwithstanding the difficulty of finding striking examples, we can yet declare on the basis of the scant material that was collected, that it is an irrefutable fact that unnecessary checks occur in scientific research and its application.

What is the relation between the "material lack of freedom" as it has been discussed in the foregoing in connection with "frustration", and the "formal lack of freedom", which on account of its spectacular character draws most attention? May one say that materially speaking the freedom of science in our Western world has not been completely realised, but that formally it is fully guaranteed? In reality things are different. Material and formal lack of freedom merge into each other by degrees. The forms of frustration that have been discussed in the foregoing in many cases boiled down to formal lack of freedom for the research-worker. When such a worker before entering the service of some institution or industry is compelled to sign a contract to the effect that he will keep the results of his research secret, even for his colleagues; when he is dependent on the consent of the management every time he wishes to publish; when the choice of his research-subject and the practicability of his results are exclusively at the mercy of the opinion of an organisation whose judgement is led by other than scientific standards, then formal freedom is completely absent. A very large part of the scientific research-workers is facing this situation at the moment. And even though in theory one might say that these people are free n o t to enter a contract that binds them hand and foot, yet in practice this would mean for many of them, even if they were to succeed in finding some other job, that they more often than not would have to give up all idea of scientific research.

This state of affairs can be imputed to the fact that science is being more and more institutionalised and is pursued under sponsorship of government agencies and commercial organisations. It is natural that these organisations try to keep control of the direction the research is to take, and that they try to reserve to themselves the decision of application or non-application. Together with the disappearance of the independent economic position of the inventor and his absorption into big laboratories, his mental freedom is fast disappearing too. Whoever holds the pursestrings, reserves to himself the right not only of determining on what the money shall be spent, but he will also try to stop "dangerous" utterances of those who are economically in his power.

The fact that the problem of "Man's Right to Knowledge and the Free Use Thereof" was given a prominent position in the celebrations for the Bicentennial Anniversary of Columbia University may certainly be taken to be an expression of anxiety about the increasing impairment of freedom in science, also in the United States themselves.

The a b s o l u t e ideological and economic freedom of the man of science seems more and more an abstraction belonging to the 18th and 19th centuries, like so many other liberties; at most they may still find a stronghold in the academic world and even that is no longer safe either. This is due to the fact that the university does not remain uninfluenced by the social system. This is especially so when universities accept grants for their research from industry, or if they are absorbed into researchwork for military purposes. "Cases could be cited", as a professor of Rochester University said "of research sponsored by research-agencies, where graduate students and others must be cleared by the Federal Bureau of Investigation, where they are required to work behind closed doors through which no one can enter, who is not cleared and where they are unable to discuss the details of their research with fellow graduate students and with faculty-members, who are not admitted to the sanctum sanctorum". [1]). The writer also pointed out the danger of research grants by industry to universities. Nevertheless the urge towards independence in university circles together with the realisation of all these dangers, including that to formal freedom, is still a great social force.

While in a certain sense material and fromal freedom are connected, as the one gradually merges into the other, yet on the other hand one might well ask oneself whether a certain restriction of freedom is not inherent in the increasing realisation of the value of science for man and society. A government that looks upon scientific and technological development as a matter of primary importance for the national community, will not be indifferent to the direction the research is taking nor to the manner of application of the discovered results. In the same way that an industrial undertaking reserves to itself a great amount of influence on the work going on in its laboratories, so shall a government that attaches a great amount of value to scientific research reserve to itself the necessary influence on the choice of objects to be examined, on the choice of the research-worker who is to be appointed and on the applications that are to be given practical effect. Any government can already do so in large measure by granting liberal means to certain inquiries and by withholding them from others.

The chairman of the Academy of Sciences in the Soviet Union Professor Nesmeianow in a recent report characterised "the strategy of scientific progress in a socialist society" thus:

"Scientific staffs must be directed towards solving the main problems exactly and precisely. By bringing science in every way closer to the practice of building communism, to industry and to

[1]) Albert W. Noyes, "Is sponsored research a danger to the academic tradition", Proceedings of the Association of Graduate Schools in the Association of American Univ., Oct. 1950, quoted by Stern, l. c., p. 117.

production, by absorbing the experience of industry and of the innovators in production, we must solve the most important theoretical problems of science" [1]).

Seen in this light absolute m a t e r i a l freedom proves to be an illusion both for science and for research-workers. As soon as science is housed in institutes, those who are in authority – whether they live under a capitalist régime or some other régime– will exercise their influence on the course the research is taking, as also on the possibilities for application, even if only by being more or less liberal with grants. The problem here does not lie in the question about freedom or the lack of it in the traditional meaning, but in the question as to how the best guarantees can be created for a maximum realisation of scientific progress on behalf of aims that have been instituted by society itself. So it is not in restriction in the usual sense, but as has been put forward before, in the restriction expressing itself as f r u s t r a t i o n that the gravest problem is to be found.

Those who took the initiative for this congress rightly made their guiding principle "Man's Right to Knowledge". This right should not be taken exclusively in its individual sense as being the formal right of each person separately for gaining knowledge, it should also be taken in a collective sense as the r i g h t of m a n k i n d to reap the full benefits of science.

Things become different, however, when those who ultimately have to decide, start arrogating to themselves the right of judging about the scientific merits of the work performed by research-workers and scholars. The concept "scientific progress" is the diametrical opposite of any sort of acceptance on authority. However great the temptation may be for some government to extend its authority over the contents of theories and systems, even though its aim may be the maximum of useful effect for society, yet by so doing it would only cut itself to the quick. By doing this it would do away with one frustration merely, to put another in its place. Every scientific renovation arose in the past out of resistance against theories clothed with authority. On this point the scientific world should with all its energy champion freedom and more in particular freedom in the formal sense. It will have to

[1]) Cf. J. D. Bernal, "Where science serves peace", in "Masses and mainstream", May 1954, p. 31 ff.

fight a g a i n s t the restrictions of freedom of the mind, a g a i n s t keeping scientific results secret, but also a g a i n s t material lack of freedom in so far as it forms a hindrance to the public circulation of the scientific knowledge that has been gained.

In every form of society it remains the task of the man of science to defend the profoundest scientific values against the established powers, whether this be a private employer or a state. To these values, besides his right to learn to know the truth and inform others of it, belongs the right of differing in opinion from the powers that be.

DEBATE ABOUT PROF. W. F. WERTHEIM'S PRE-ADVICE ON "ASPECTS OF FREEDOM AND RESTRICTION IN SCIENCE"

by M. G. PLATTEL O.P.

What I appreciated in Prof. Wertheim's introduction is that, according to the pre-adviser, within the realm of science only the criterion of true or false should count, and that tuth should not be violated for the sake of considerations of utility.

In reading the pre-advice, however, I believe a certain one – sidedness is to be noted. This need not be onderstood to be disparaging if it is based on the fact that the speaker only wished to treat a certain aspect of the problem.

It seems to me, however, that the pre-adviser does not proclaim science to be a good but the highest good. Moreover he speaks of science too much in the abstract and realizes too little, I believe, that we are concerned with science that is practised for and by human beings. When a country therefore spends a great deal of money on purposes of defence and has to curtail its expenditure on strickly scientific research this may sometimes be entirely justified, because there other human values than those of the knowledge of science only.

Prof. Wertheim claims absolute freedom of scientific investigation, and this scientific research is clearly confined in the pre-advice, as appears from the examples, to that of positive science, the science of fact. The speaker strongly creates the impression that he sees the highest good in an ever greater knowledge of the inter-relation of facts, to which all values must be sacrificed, thus on p. 100, as an example of frustration in science, are indicated the impediments of: ideological values. On p. 100 these impediments are rejected together with others in the summing up of the final conclusion: "In the foregoing, attention has chiefly been paid to cases of frustration of so extreme a nature, that there could hardly be any doubt among men of science as to the correctness of the label".

Now I should like to put this question to the speaker: does he wish there to be no single impediment of an ideological nature in the pursuit of science and move especially in that of positive science? May experiment be made on human beings if need be, which might and in death, so long as the knowledge of science could be furthered in that way? The pre-advice might quite easily result in a confirmation of this question because ideological impediments are too easily rejected and positive science is really raised to the positions of the highest good.

The speaker leaves the scientist's views of life out of account. The pursuit of the positive sciences should not be detached from a view of life, however, and is bound up with a hierarchic scale of values. The speaker can not deny that in defending the freedom of science he already expresses an appraisement, leaving the realm of positive science behind him. To over-accentuate the good of the pursuit of science may therefore in volve a false judgement if it means disregarding the hierarchic order of the precedence of human values.

PROF. WERTHEIM'S ANSWER TO PROF. PLATTEL

It seems to me that Prof. Plattel's remarks are largely due to misunderstanding. First and foremost I nowhere in my pre-advice "proclaimed science to be the highest good". On p. 90 I am perfectly content to establish that science "represents a positive value". That there are limits to the power and value of science is not a point in doubt; but these limits are not under discussion to-day, but were so yesterday.

And where did I ever say that ever greater knowledge is "the highest good to which all values must be sacrificed?" On the contrary, in my pre-advice I point out on p. 99 that every pro-nouncement on a deficit in science contains an appraisement based on a certain view of human welfare and connected with some ideology or view of society. When I state that ideological restrictions may act as deterrents to the progress of science, then that by no means is as much as to say that the ideology must necessarily yield. The decision reached depends, as was already said, to a large degree on a person's subjective view of society.

The esteemed debater has consequently read me wrongly if he believes that the sentence on p. 100 to the effect that "in the fore-going attention has chiefly been paid to cases of frustration of so extreme a nature, etc." refers to all the possible impediments of an economic, ideological, etc. nature. On the contrary with this I only wish to say that the *concrete examples* I quoted in the fore-going were selected in such a way that but little difference of op-inion might be expected; but this implies that in numerous *other* cases there may be much greater reason for doubting whether it was possible to speak of "frustration".

What surprises me most is the debater's criticism to the effect that I am supposed only to have spoken of science "in the abstract" instead of realising that what we are concerned with is science practised for and by human beings. On the contrary, my whole pre-advice is entirely governed by the idea that science should first and last be judged according to its concrete value for man-kind. On the strength of my view of the general welfare of mankind I condemned the great stress laid upon purposes of military defence in scientific research.

The esteemed debater has again not read me properly when he

says that I claim "absolute freedom" for scientific research. On p. 102 and following I even try to show that absolute freedom for the man of science is illusory. If the debater had based his criticism on the text of my pre-advice rather than on impressions and supposed aims, then misunderstandings like these I am quite sure would have been avoided.

To and with, however, I should like to mention one point where Prof. Plattel and I probably do differ fundamentally. I believe that science should certainly not prevail over ideological restrictions under all circumstances. But on the other hand I am convinced that the continual progress of science will in the long run cause the ideological restrictions one after the other to yield. Ideologies, if they wish to keep sway over the minds of people, will have to adopt themselves again and again to what has meanwhile, scientifically speaking, become common property.

DEBATE ABOUT PROF. W. F. WERTHEIM'S PRE-ADVICE ON "ASPECTS OF FREEDOM AND RESTRICTION IN SCIENCE"

by G. GONGGRIJP

When Wertheim says that mankind owes all its culture to the process of technical and scientific progress, then that is taking things in much too narrow a sense! Song, dance play and sport are all aspects of culture. An institution like the weekly day of of rest is of very great importance from a cultural point of view and not only in an economic and social respect. However, I must drop this subject.

Science may be called restricted, Wertheims says, when its progress and self-realization are hindered by all sorts of factors. To my mind this is an unfortunate formulation of restriction in science, a concept that we must duly distinguish from hindering. Science is restricted by the fundamental rules that hold good for it, or, which comes to the same thing, for its students.

Wertheim's pre-advice is not really concerned with restrictions but as appears clearly and as he himself has said, is about the frustrations in science. Thus, according to Wertheim, the possibilities we possess in science are often not properly exploited, used, realized. To my mind it is only natural that this is so. We are fallible creatures after all. The question is whether this can be called an evil, a problem of serious dimensions.

According to Wertheim our social structure brings along with it a one-sided technical and scientific development, a proposition which to my mind remains entirely unproven in his pre-advice. This one-sidedness according to Wertheim shows most clearly in the following figures which were taken from official publications in the U.S.

Out of the money spent by the Federal Government in 1954 about 90% was expended on research and development for military purposes. The proportion – or trather the misproportion – is clearly manifest when one knows that the research financed by the Federal Government for non-military purposes was increased a little more than four times in the period 1940–1952, while its expenditure for military research in that same period was increased forty-four times.

What do these figures prove with respect to the frustration of science? Nothing! The research done for the sake of defence will in the first place in many instances quite possibly be of benefit to science. And above all I am extremely glad that such large sums are expended on military research! Not only is this one of the conditions that must be fulfilled in order to allow us to go on living as free human beings, but it is also a proviso for being able to maintain the freedom of science. *In order to maintain our own freedom and that of science we must be armed to the teeth!*

It is moreover important, thus Wertheim continues that about 94% of the Federal expenditure went to applied "research and development", and only 6% to fundamental research.Does it appear from this, as Wertheim wants to have it, that scientific development in our Western world has become onesided and unbalanced?

To my idea not at all. Our chairman or a philosopher or a historian will probably receive a very modest sum for the requirements of their research and work, whereas a phycisist who teaches experimental physics, a technician and an astronomer may perhaps get large sums apportioned to them. This by no means proves one-sidedness or lack of balance in scientific development, much less frustration of science.

It is Wertheim's opinion – this appeared to me in talks I had with him – that speaking generally it may be said of the nineteenth century that the employers have fulfilled that part of their task which consists in selecting from among the technical discoveries those that are both useful and profitable from an economic point of view. It is also his opinion, however, that this favourable view of the activities of the employers is no longer justified in our own century. He gives some examples, like that of F. A. Philips who warns against making long-life lamps which might injure sales. He calls this a clear example of frustration of scientific progress through an economic factor. I cannot see this. We have scientific progress, or at least technical progress, when we learn to make lamps, whether of long or short life, that it was not possible to make before. Whether it is in the interests of business to make products of a longer or shorter durability will in most cases be possible to calculate with a fairly large amount of certainty. Whether the interests of the employers and the com-

munity coincide can really only be determined in each case separately, and then only with the greatest difficulty because as
Wertheim also admits, what people so easely call the interests of
society are often so difficult to ascertain.

What to my mind *can* be said is that the general interests of
industry and the interests of society in general will coincide if
there is free, competitive production for the market. That employers in this age too, generally speaking, still serve progress by
their activities cannot to my idea be doubted. It still always appears that production generally is served by free activity on the
part of the employers. That continual supervision by the authorities, enlightenment of the public, vigilance against misuse of
power, etc., will always be necessary goes without saying.

One of the differences between the nineteenth century and our
own time in my view is that in the nineteenth century the rise in
production was of almost predominant importance. Now too of
course increase in productivity is still of great importance. But
the need for stabilization and continuance of the industry is much
more powerful now-a-days, I think, than in the previous century
when there was poverty on a so much larger scale. From Wertheim's pre-advice I get the impression that he has not kept this
point sufficiently in view.

That insufficient financial means may hamper scientific research is self-evident. I should hardly dare to call it frustration of
science however.

That resistance against what is new may impede the progress
and diffusion of knowledge goes without saying. Only one might
observe – to my idea an extremely important sociological point –
that the conservatism existing in the human heart has a very important function to fulfill, namely to make order possible. If we
were not generally inclined to repeat our daily activities, no kind
of order would be possible in the world. By this I least of all wish
to pronounce a damning opinion on progress.

I also am more optimistic about the consequences attendant
on keeping the results of scientific investigation secret than Wertheim. Generally speaking it will not be possible to keep new
scientific views that are at all worth while secret for long.

Wertheim's pre-advice has not convinced me that we can speak
of frustration of science as if it were an important fact, a problem.

I have gratefully noted that Wertheim ends by stating explicitly that we must defend scientific values against everyone and everything, whether it be an individual employer or a state, and that it is our inalienable right to differ in opinion from whom we will.

If we wish to have an example of serious frustration in science, in the United States, then I should wish to mention the following. Frustration was there carried on by a man of science of international renown, by Owen Lattimore.

The Chairman: the time is up.

PROF. WERTHEIM'S ANSWER TO PROF. GONGGRIJP

When I state that mankind owes all its cultural property to the process of technical and scientific progress then I am very far from wishing to exclude the contribution the artist makes to our cultural inheritance. But art in turn avails itself of technics and science, from pottery making to the most exquisite manifestations of literature and music. Only via technology could all the various aspects of our civilisation be realised, from making a fire to the most modern orchestras. All I had in view therefore with what I said was to lay down that our civilisation in all its various aspects presupposes technical and scientific progress as its basis, and I believe that taken in this sense I can stand by every word I said.

The restrictions in science through its fundamental rules have been fully discussed yesterday in the cadre of the boundaries of science. By the limitations of science as opposed to its freedom, the subject under discussion to-day, is meant restriction through factors that lie outside science as such, the external factors. If the esteemed debater has any objections against this terminology then he should address himself to the organisers of this congress, not to me.

As regards frustrations, colleague Gonggrijp's argument really comes down to this that he denies that in society as we know it there are any fundamental deficiencies in the field of scientific activities. From the fact that a given process is not applied the debater concludes that the process *for that reason* does not pay, and that therefore there is no sense in complaining about this non-application. But it was my intention to warn against this quietism, against accepting the present state of affairs as a matter of course. Colleague Gonggrijp's attitude, as it is expressed in his whole opposition has too much about it of the "tout est pour le mieux dans le meilleur des mondes possibles". An appeal to human fallibility keeps us from making a proper inquiry into where the shoe pinches. All my examples of frustration serve to make clear that there is a big gulf in the present state of affairs between what science actually achieves in society and what it might obtain.

This attitude of acquiescence in colleague Gonggrijp comes out very clearly in the thankfulness he expresses that America is armed to the teeth. I should like to ask him whether his gladness goes so far that he rejoices that there is a world situation in which in his opinion such armament is necessary. If not, then I should like to maintain that this world situation itself is at the least an obvious cause of frustration in science.

And further I should like to point out that there is a large part of the world to wit the whole conglomeration of under-developed countries, where the frustration of science goes so far that the peoples in question are almost wholly devoid of the blessings of science. This situation is aggravated by the fact that the present armament-race leads to an ever greater stress being laid on the production of strategic raw materials, which means that it is impossible for the peoples in question to raise the standard of living in any way. "Stabilisation" judged superficially may be more or less acceptable to society as we know it to-day, but for the peoples inhabiting the underdeveloped areas such an economic policy would be an unbearable sentence condemning them to go on living amidst the greatest poverty.

I also consider the idea that processes cannot be kept secret long, much too optimistic. The secret that the Stradivari took to their graves with them seems to me a very good example of the reverse. Once again we see what *has* been applied but we have no means of finding out how much useful knowledge is withheld from us through it being kept secret.

Taken as a whole, this in contrast to Gonggrijp, I am very far from being convinced that the interests of the employers regularly coincide with those of the community. That the frustration of science is a very important problem for present-day society seems to me indisputable for whoever is critical of the existing order of things.

Finally I should like to repeat, that when one is speaking of man's right to knowledge, one should not take it individually only, but there should be question of the right of all mankind to benefit from science. I hope that this Congress is able to agree with this view and interpretation.

THE CONSEQUENCES OF FREEDOM AND RESTRICTION IN MEDICAL SCIENCE

by H. W. JULIUS

It is the task of Medicine, to act as the protector of man's existence. It tries to prolong life, to alleviate suffering and to further health. The time that man sought to do this through intuition, tradition and good faith, is past. Medicine in all its branches is fully alive to all the possibilities that science has offered; without scientific study medicine would be, if not powerless, than at least very much less in a position to take the active part it does, than is the case to-day. On medicine, over and above the very varied tasks of all other branches of science and learning, rests the protection of our *own* species. Medicine touches the ego, directly and vitally. It's object and it's practitioner are in essence indissolubly united. This characterizes its special position.

Medicine has another characteristic, namely that it is concerned with what is diseased and defective, with pathological states, whilst outside medicine we do not come across the study of what is defective [1]).

On account of these special characteristics, within the framework of the pursuit of science as a whole, the two antitheses Freedom and Restriction will also have a special significance in the science of medicine. For where the ego is involved it is very difficult to judge its problems at all objectively, the more so because it is concerned with what is diseased, which arouses an instinctive feeling of aversion and which conjures up the desire laden with emotion to banish it.

In a society that feels it necessary to raise the problem of Freedom and Restriction in Science as being a fit subject for discussion in an ambitious whole, in a society I say, in which these concepts have apparently become confused, there it will not be so easy to bring to clarity the line which Medicine above all other branches of knowledge, has been apportioned to seek.

What follows is only an attempt to arrive at a definition of

---

[1]) Except in the veterinary sciences that is to say, and in the very young science of phytopathology! These two branches of science in their very nature are derivatives of human medicine; their problematics are thus linked with those of medicine.

concepts, not for Medicine only, but also for its object and protagonist, man. Nobody, I hope, will expect me to give a solution.

Science – or rather whoever is engaged in the pursuit of science – feels the need of a sense of freedom. In and through science and knowledge we strive to fathom truth. It is irrelevant whether in doing so one ever really comes to know truth. It is even unimportant in how far one can fathom it, or how one experiences the knowledge of this supposed truth. It does not make the desire for freedom any different, and it certainly does not make it any the less strong. On the contrary; the craving for approaching truth brooks no curtailment. He who is engaged in the pursuit of science, the researcher therefore, requires a clear field. It is his right to approach truth in the manner that seems fit to him, to arrange his thoughts as it please him, to seek out his object, and to carry out his experiments, or his method of working according to his own discretion. The man of science is absolute in this respect. *Any and every* sort of restriction is felt by him to be a hindrance, an intolerable deprivation of freedom and with regard to this he is positively aggressive; he must be even, because ... freedom without a readiness to fight in order to protect it, or even to attain it, is an internal inconsistency.

Put thus, the matter is simple enough; but it is only one step from the simple to the simplification. The concepts of Freedom and Restriction should be placed against some sort of background as they are anything but absolute concepts. Only a dreamer, a fantast can imagine an unbridled freedom in the pursuit of knowledge. I shall quote a short satire by A. Standen, entitled: "What will Scientists think of Next?" [1]:

"Another invention crying aloud to be invented is a genuine gravity-removing device. A screen that cuts of gravity .... But we have seen no results yet". (Cf. H. G. Wells) [2].

After giving a description of an arithmetical machine which can much better perform in a few seconds that, for which a whole army of arithmeticians would need considerable time, he says: "It would seem entirely right by the standards of modern progress that professional translators should be rendered technologi-

---

[1] From Science and Society, a symposium. Lobond Institute, Notre Dame, Indiana, 1952.
[2] The First Man in the Moon.

cally obsolete.... And yet I don't know of any progress being made at all by any of our industrial firms in making a translating machine".

Indeed, science is linked to the properties of matter as well as to those of the mind. These links are primary, *intrinsic* and absolute. This is not so much as to say that the plane where these links are to be found is constant. On the contrary. This level changes with the times: that it is possible to cure a disease promptly with some tablets (for instance sulfanilamide) with almost absolute certainty was once an absurd wish. He who wished to believe in that "philosopher's stone" was a fool. That this wish has come true does not alter the fact that these links exist, even now, even though the plane is now different. But who shall say whether it differs much?

He who is not a fantast knows and will unabatedly acknowledge that these links exist, but... he does not experience them as such. It can even be said that they are nothing to him in the least and that they will never stand in his way to continue using his best endeavours. The difficulties consequently beset us from another quarter: I said that whoever is engaged in the pursuit of science feels the need of a *sense* of freedom. The subjective element has here been brought into discussion – on purpose – it has even become the decisive factor. With this all sorts of problems arise since the relativity of things has been introduced. We shall have to consider this more closely.

It at once appears that there are two forms of restriction of freedom, namely *intrinsic* and *extrinsic*. The most extreme form of intrinsic restriction we have just discussed. It concerned the impossibility of approaching nature otherwise than through the existing systems of openings accessible to man, let alone to force it to take on the shape of a dream. This restriction, we saw, is acceptable, unreservedly.

Against what then does the spirit of science rebel? In accepting the most extreme form of intrinsic restriction it is the best to characterize the unacceptable by its opposite, namely the most extreme extrinsic curtailment. If the reader wishes an example, the following might serve:

A certain political school of thought needs certain tenets respecting the genetic development of man in order to support and justify its tendentious aims, What genetics teach is not in agreement with this, however, so this branch of science will not be able to find full expression under this system, or worse, it may not make its findings public, or still worse, its findings will have to be tampered with through incorrect rendering and (or) interpretation.

We now need not hesitate to allege: this is in such flagrant conflict with the very nature of science that there is no other problem to be discussed here than "the right of the strongest". Whoever wishes to be stronger than nature will only be able to sojourn in that dream for a limited space of time.

Therefore we can accept intrinsic restriction in science, but extrinsic restriction we reject. Put thus strictly the matter is clear enough; and so we have answered the question we put .... were it not for the fact that the two following questions recall us to reality.

Where does the boundary-line between intrinsic and extrinsic restriction lie? and,

Are there no border-line cases near or on that boundary-line, where the two opposites approach each other so nearly that confusion of ideas is (almost?) unavoidable? Let us then try to find the boundary between acceptable intrinsic restriction and unacceptable extrinsic restriction, and let us see what is happening in the territory where the boundary lies. To restrict myself to medicine only is not possible. Examples can be taken from medicine and peculiarities can be pointed out, but the internal connection of all branches of knowledge is so intens that there can only be question of a general stress. I hope therefore that it will be permitted to me to make a wider grasp.

At the beginning of the cycle of all scientific research is *the idea*, the hypothesis, or if you like the "conception". Then follows the making of a *plan*, first still only vaguely, then more in detail, and all the time capable of continual revision and adaption. If the plan is *to be carried out*, then the *means* to that must be devised and realized. This means that the *object* must be determined and *the observation apparatus* in its widest sense devised and put into working order. Then comes the moment at which both are confronted with each other under the searching eyes of the investigator. Once the research work has begun, it will appear whether the question put to nature was right; perhaps it will have to be altered, then a shorter or longer series of successive *observations* that correct each other will unroll itself, until the *answer* to the question that was put in the first place appears as the *result*, or perhaps a quite different answer that fits a question that had not originally been put. This result may be strict or shaky, it may lead to the enrichment of knowledge, of knowing, or it may increase the problematical side of the whole thing. That in the end does not matter. And then of course comes the conclusion, accompanied by self-criticism and criticism and then ... the disclosure, the *publication*. Making public the results arrived at in the pursuit of knowledge has become a very important phase nowadays, and it means that the investigation has got into a fresh state of aggregation. Just as man at birth becomes a sharer in a society, so does a piece of research work at the moment of publication. That it is a chill wind that blows here, may not be every one's opinion as it most certainly is the author's.

But this is not all. There is another phase, which is often forgotten or neglected; for our subject it is very important, however. What has been discovered, the result, the "knowledge" will take its course. Earlier or later the product of science will have to endure the rays of searchlights of a lesser or greater brightness, it it will cut out its own career, it will become: *Public Property*. This may mean: a reason for renewed research. Or: revision of opinion; or: Application. In the case of the matter being hushed up and the search-light never getting a chance, the end is quickly reached, in the other case the aftereffect may be very considerable; a long and radiant life. But whatever the case, not until this phase is reached can we speak of the last one; only at the end of *this* phase is the career of the investigation closed. A new idea is necessary in the next cycle to raise the plane that has been reached.

1. *The idea.* Over 15 years ago I remember writing: "creative work can only then be performed, if the research worker is an artist first of all".[1]). To-day I am convinced of the correctness of that view, even more than I was at that time. The antithesis science and art is for me only a seeming one. No more than the

---

[1]) See also W. I. B. Beveridge "The art of scientific investigation". Melbourne, London, Toronto 1950.

artist can work without inspiration, no more can the scientific investigator do without those sublime moments, which are independent of material bonds. Just as the artist propagates his ideas with the help of matter, so does the investigator also carry out his ideas with the means that science offers. That the way things are performed in art and in science differ, is immaterial. In science just as much as in art the artistic conception lies at the root. The analogy holds good still further. The inner fire of the artist and the scholar is equally strong and equally absolute. Their belief in the creations of their minds is equally firm, and the shape they give to what is in their minds is equally harmonious, or, if you will, disharmonious. The *truly* elect among them are equally rare with both. And, the thing that matters, they both have an equally burning desire to freely realize their conceptions. There they can follow their own bent freely, without interference from the outside world: free from extrinsic bonds, while accepting the intrinsic ones.

What marble is to the sculptor to which he gives shape in his imagination and by whose natural construction he is bound, that to the medical student is the object, man or some other living creature from whom he hopes to elicit the answer-to-his-question (i.e. *his* shape or form). That they are necessarily bound does not trouble them, it is included in the vital necessity of giving shape to their ideas, which need the resistance of the object to make their performance complete. Both the artist and the investigator feel the need of helping man, of presenting him with gifts. This is done a little more directly perhaps in the case of the medical student, but it is a small quantitative difference. He – the medical student – who might feel it to be an essential difference, would then make the same mistake that is made by so many artists! This then is as it should be. Because whoever does not feel free while experiencing some scientific idea does not accept the intrinsic bondage and is very near to the fantast, or may even be considered as such.

Nevertheless there is another conflict concealed here. Is it after all possible to give one's ideas free reign at all times? There are fashions in science, certainly in medicine, there is a bias and dogma, one man even speaks of scientific dogmatics by means of which the shaping of scientific ideas is conducted into certain channels for long periods. The restrictions of ideas that is thus increased is, if we look at it objectively, extrinsic, but the curious fact arises that this is not observed either. If at a given moment when the idea comes to birth, the incorrectness of some dogma is discovered, then whoever does so is for an instant free from this restriction; if on the other hand the investigator follows the dogma, then he does not feel his bondage, at most it can be recognized by him who has freed himself in his own time scheme, or at a much later period. This makes no difference to the requirement that the person engaged in the pursuit of science should *feel* free. In any case we have learnt that already at the very earliest phase of scientific investigation the concepts that matter, the intrinsic and extrinsic restrictions are relative, although in this case they are still recognizable. But it already becomes more difficult at the next phase.

2. *The plan and its execution* are so closely bound up together and at the same time so bound to reality, that any demarcation line between these two phases would be artificial. Although the plan in its narrower sense may be immersed in the same sense of freedom in which tne ideas arose, yet it is so absolutely bound to the means of execution – at least if one wishes to remain a realist– that this bondage might almost be called intrinsic and therefore acceptable, were it not that it appears on a closer view that forces from outside have really come into play. These lead us to suppose that restrictions of a non-scientific character, in fact of a human arbitrariness, in other words forces of an extrinsic character, have become strongly co-determinant. Indeed at this stage of the investigation one may already be faced with considerable extrinsic restrictions in freedom, which are experienced as such and against which the feeling of wishing to remain unfettered does indeed rebel. In the very first manoeuvring, where one was still engaged in arranging the succession of questions to be put with the expected answers, there the restrictions may still be considered intrinsic. But as soon as the object and the appliances come up for discussion, circumstances arise which to my mind are of such a character that they helped to give rise to the desire to consider the subject of this congress.

The object of medicine, as has already been said, is man, no more, but also no less. As man is the object one can only approach it with the greatest reserve. The aim is to render help; there may of course be no question of harm accrueing as a result of investigations; for what counts in the first place is the alleviation of suffering. But if one wishes to attain to this, then it must be tackled causally, and the humane deed simply cannot be performed unless one knows how to fight the disease. To that and it is necessary nor only to know the normal constellation, but also the possible deviations in the normal completion of the functions, as well as the causes that give rise to the deviations. One can sit and observe. It makes a little the wiser, but not much. One soon feels the desire to look deeper and to understand better. One wishes to approach one's object much more closely and to unmask its riddles. One is then faced with the problem of how to do this without causing harm. And here we come up against the problem: what is harm and where does it begin? Man has a deep-seated feeling that it is unpermitted to encroach upon man, so much so that some peoples experience the touching of the body for examination as a violation. One is only allowed to talk about it, every attempt at prying into matter is "tabu". This idea was very deep-seated in our own history too. It is only about four centuries or less ago, that he who ventured to perform an autopsy was liable to be persecuted for violating corpses (Vesalius). That the doctor was allowed to see the whole process of child-birth instead of only rendering help by feeling under the sheet is of a still more recent date. The freedom medicine to-day enjoys has been hard-won. And in many cases it may be said that the outsider still regards them with disapproval. To our feeling, which is attuned to the standards prevailing to-day, science was very much tied where man was its object. What we now should like to know is whether this restriction of freedom was intrinsic or extrinsic.

If we look at the impossibility of performing an autopsy with modern eyes, then we call such a restriction a form of human error of the past; pre-eminently extrinsic. Doctors in those days, however, – and then I do not mean the pioneers who did secretly what could not be done openly – considered this restriction as a natural corollary of their branch of science, as self-evident, as... intrinsic and therefore worthy of acceptance.

People will object that I have made a mistake in thought since I have been making use of two concepts for intrinsic, which have been brought together on the strength of the similarity of their acceptability. I agree. But I made this mistake on purpose, because it is a confusion of ideas that is not rare, and because it is still entirely of current interest; it is even the actual basis for this gathering.

Indeed we have a striking example of this in medicine; I presently hope to be able to show that the confusion of which I spoke is of current interest, and what its effects are. In the other branches of knowledge as we shall presently see, the pitfall of this confusion does not appear until a very much later stage.

Only in medicine is the object tied in the manner meant above. "The physicist bombards and destroys his atom nuclei without the slightest compassion; the chemist makes his molecules pair incestuously or separates them forcibly without any consideration". [1]). But that which has been entrusted to the custody of medicine may only be surrounded with care.

But what is there to quarrel with, when only the inanimate part of man is left, to study by its means what stood in the way of the continuation of the association of body and soul? The reserve regarding post-mortems has indeed been removed, if not in all people, then at least in many; the gain this has meant for the protection of life is incalculably great, every day still. We have relinquished our former bondage – quite apart from how far man has progressed in wresting himself free from prejudice. But with *animate man* it is a different thing Although we know that there are innumerable problems that might be cleared up by studying living man, whether diseased or healthy, it in such flagrant conflict with the task of medicine, that we must condemn it. This has meant the restriction of science. And on account of our marked readiness to accept this restriction, the semblance has been created that it is intrinsic. Why is this? The moral bondage of the medical student has been identified with the bondage of science to nature. And this is disastrous, in both directions. Thereby the possibility has been given scope to break through the moral bondage with motives in favour of scientific freedom, and people sometimes try to justify the curtailment of

---

[1]) F. F. Rochat, in "Zesde interfacultaire leergang". Academiejaar 1947–1948. (Sixth inter-faculty course. Academic year 1947–1948). Groningen-Batavia.

scientific research by disguising the true motive in the garments of rational argument. To give an example of both:

The first is the must striking: one need but think of the revolting experiments on people occasioned by an absolutist régime – only too easily accepted by a multitude out of control – whilst in actual fact morality has been let go of, the mode of procedure is justified on the grounds of scientific gain, or possible fancied gain. It is even argued that the aim of science can only be served by absolute freedom, that it is unlawful to impose restraints on science when it is seeking truth. A much stronger sense of moral responsibility is necessary than the unstable multitude possesses to discover what confusion of ideas has taken place. It is curious though, how science with medicine in the forefront, fights for scientific freedom with as great ardour as it does for moral restriction, This is a gladdening sign. But because of the simular mentality by which these two aspects are experienced, confusion threatens to creep in all the same, unless one sees what one is doing.

Now the opposite: this is much more difficult because it is even less readily recognized. For an appeal to moral restriction is easily accepted as being intrinsic. Now then. The restriction that what is observed in an experiment on animals cannot without more ado be considered as giving insight into the processes of man, that is truly intrinsic and therefore accepted. All this is grist, that comes to the mill of the anti-vivisectionists who disparage the scientific gain flowing from experiments on animals with all their might, and who point to the sometimes most unfortunate mistakes, that were the result. Here again therefore we find essentially the same misapprehension. On the grounds of moral motives it is possible to reject experimentation with animals – whether this can be maintained or not I wish to leave an open question – but it is not possible to do so on the grounds of the danger that threatens when the gulf lying between the reactions of men and animals is rashly bridged over. The latter is a scientific problem, an intrinsic restriction; whereas putting a curb on vivisection is extrinsic and therefore, from the point of view of the pursuit of knowledge, unacceptable. What one thinks of all this is open to discussion, but it is a confusion of ideas if one imposes restraints on vivisection on the grounds of scientific limitation. And yet this happens.

I might continue with confusions of this sort; they are numerous. But I shall be content once more to warn against confusion of a deep-seated sense of responsibility that lies in the feelings of the (medical) investigator with respect to his object, with the intrinsic restriction as to the degree wherein he can approach his object scientifically. They are two *separate* fields.

It would be tempting to consider the shifting of the frontiers of moral standards more closely. I shall only point out that the old prohibition regarding autopsies has made way for lawful experiments on living man, provided that he has voluntarily put himself at the disposal of experimentation at a moment that his decision was made in absolute freedom. One may note even that hero-worship [1]) has taken the place of the former condemnation. And also that experimentation on man becomes ever bolder, both as to extent and frequency as well as to publicity. But we cannot enter more fully on the moral aspects of this problem. Because in this phase there are many more restrictions which – as best as we may – we must analyse, especially because to the sense of freedom the acceptability diverges so greatly.

For instance what about the material limitation of means required by an investigation? When one makes a plan for which large sums of money are needed for carrying it out, then there is of course somewhere a limit to the possibilities. This is reasonable and on the strength of this reasonableness it can also be accepted. It is so self-evident that there *must* be a limit somewhere, that one immediately realises this in making the plan. The financial limit therefore already makes its influence felt in the making of the plan, just as much as the idea is determined by the intrinsic bondage. And it is reason which makes the restrictions which are actually extrinsic, imposed by money, acceptable and makes them seem intrinsic. However, this is how it is only when very large sums are involved. In the case of smaller sums the restrictions thus imposed *are* felt and they are therefore not acceptable, because they are clearly extrinsic. The battle is waged in the (wide) borderland. Via money the non-scientific side of society has an enormous influence on the pursuit of science.

---

[1]) I refer to the international Walter Reed Society, to which anyone can be admitted and bear the signs if he has put himself at the disposal of experimentation, and provided that the experimenting brought with it some risk above a minimum standard.

Perhaps it does not sound very kind, nor grateful, nor does it evince any great delicacy when I openly declare, that the scientist, under the pressure exercised by the society in which we live, is really venal. The scientist is for sale on account of the position of his family, the equipment of his laboratory, the accompanying position in society and his chances of scientific fame, and freedom in science is impaired by it, whichever way one turns it. But ... it has the semblance of being acceptable. For it is so absolutely inherent to the realisation of the plan that in many cases it is accepted with the ease of the truly intrinsic restrictions. Always excepting the very great spirits among those who practice the (medical) sciences, the elect. But they cannot determine by themselves the evolution of science, notwithstanding their superhuman resources. The aristocracy of the mind has always had much less influence than they themselves thought or the world feared. Outside that circle science has to bow its head. We simply *have* to be realists if we indeed wish to attain what is practically attainable. Immersing oneself in wishful thinking would not carry us any further.

Indeed, money alone is enough to impair scientific freedom. And oh, vicissitudes of fortune: as much in the sense of arresting (certain) developments as in furthering progress (in a certain field).

In this way then society has the trend of scientific investigation entirely at its mercy, however much we might wish it otherwise. The (collective) conscience of scientists and scholars is not free of guilt in this respect. For do they themselves not also have double standards. Industry is sometimes accused of trampling the freedom of science under foot [1]) with its (opportunist) research organisations, that it has turned science into a factory of results, and especially of results that can be applied. But meanwhile the men of science co-operate heartily in "organizing" scientific research. Are there not all over the world innumerable and sometimes very powerful foundations, funds, organisations, of private origin or not as the case may be, that wish to further science, either generally or in a certain direction, which institutes

---

[1]) It must be mentioned that also words of appreciation are being expressed. This appreciation even amounts to jealousy; but this detracts nothing from what is written here!

meet with the whole-hearted co-operation of the scientific head pieces. They advise about the investigations that are to be undertaken or not as the case may be, and they say where the limit is to be materially, of what is to be spent on the research, and also whether this or that investigator is capable of carrying out a given "commission". I know that it is better to be directed by a body of scientific headpieces, because they know how science is accomplished, than by a body that is obliged to be opportunist, as for instance a purely governmental organ. But the fundamental lack of freedom is no less on that account. The advantage of the first case principally lies in the ability to respect the *feeling* of freedom. A favour thus allowed is not felt to be one, and that is fair.

Our subject is becoming more and more dangerous. The man of science has already passed under a number of yokes. Fortunately his feeling for freedom has not been too greatly injured!

Let us now see what becomes of scientific freedom, when the investigation passes from the initial stage of preparation into that of execution: in other words the

3. *Confrontation of the object with the means of observation.* [1]

This is the most important stage in the investigation. It decides everything, from the usefulness of the idea to the effectiveness of the plan and the estimated means. At this stage, if necessary, the idea will be revised, the manner of approaching the object will have to be adapted and the recording will have to be selected. Here imagination will have to be silenced, and the severest check and self-criticism will have to be maintained. In short, everything here is under the absolute sway of reality. Here the investigator is bound with every fibre to the situation he has himself created, he may not assert himself, he must be strictly instrumental only. And yet ... he feels entirely free! He is able to enjoy intensely the excitement of observation, he can give his thoughts free rein because reality will soon enough recall them again. He can create with his hands, his performance becomes his own creation. He

---

[1] By this term "means of observation" I mean to convey everything that can help wrest the answer from the object of study. It is the scalpel of the Anatomist, just as much as the electronic microscope or the medicine branded with isotopes, as well as the sick man's drop of blood that is brought into the healthy animal. The term used does *not* define the idea.

has this feeling of absolute freedom because he is no fantast and because he knows that he is only intrinsically bound to his object and his appliances. The chances of an unacceptable extrinsic restriction are small. On the whole the outside world respects this (spiritual) reclusory period of the investigator. It is a sort of refugium, a period of sacred peace. Not that I mean that it is a period without cares, for instance on account of insufficient reproducibility of the experiment, minimum of requirements for the number of observations, the menace of delusion in the observation, and the – sometimes experienced as painful – limitation of one's own ability. Oh, no. That is no different from the torment the artist experiences when, in the sweat of his brow, he creates his work of art. But all this sort of bondage means to the feelings: liberation, the losing of oneself in the situation. I am optimistic about freedom at this stage in the pursuit of science. Here it is pre-eminently unassailable to extrinsic restrictions.

But no sooner has the period of observation come to an end, than this dream world comes to an end too. For now we have to proceed to

4. *The conclusion.* We shall have to give content to what has been observed. If this is in agreement with the state of science then all is easy; but if dogmas are to be attacked then one will have to become more militant, with greater or lesser enthusiasm. Within the world of science, however, respect for liberty of interpretation is still (or should I say: sometimes still) present. The question as to whether the interpretation will be confirmed by succeeding investigators is beside the point. But ... the results are not confined to the sphere of science only. There is also society. It has had them dished up for it, it has tasted of them and considers them its property. For the great majority of men and women, however, relativity is an unwelcome matter. Certainty and assuredly so when it concerns the ego, is pleasanter, and it is a thing they will not soon let go. So when the results of a piece of research work are not in agreement with the current positiveness, then the liberty of science disappears into thin air at once. Again it does not sound kind nor respectful to establish the fact that society does not accept scientific freedom when certain vulnerable points in its system are struck. As far as that is concerned it must be admitted that we have progressed

since the time that religious dogmatics put science under the ban, and relentlessly excommunicated the pioneers amongst its practitioners, or – even silenced them at the stake. However, we may not speak of more than "advanced". Does society for instance accept the medically established scientific views regarding upbringing? The moulding of the personality requires a very particular degree of freedom, that should be properly balanced by the equally forcible requirement of proper guidance. I venture to propound that to realise this is only done in words. That is of course already a great deal. But it is selfdeception for instance to contend that the knowledge regarding popular mental hygiene has also led to the acceptance of the consequences. In many respects the attitude of society in this regard is as dogmatic as in the times that post-mortems were considered to be the violation of corpses. The cloak in which dogmatics is enveloped is no longer heavy with gold brocade, however, or of a deep red colour or jet-black, with long sleeves that fall in deep folds from the elbow. It is now a jacket or overalls. But the real difference is less great than the seeming difference. One may have established scientifically that man does not attain to his full stature, unless he has acquired a proper freedom; yet society ties him down in ever tighter coils. Where else do the innumerable, the oppressive amount of difficulties in adaptation come from, if not from the fact that society is not at all willing to accept the scientific views, and that it imposes its will just like the inquisition used to consider its proper right.

In one field meanwhile there is progress to be recorded. And that is the penultimate stage of every investigation, namely:

5. *Publication.* For there is an almost unlimited possibility now-a-days to make public what one has "found" in science. Thousands of reviews are there ready to help. One has already become accustomed to astronomical figures. After all it does not matter whether one estimates the number of publications in a given scientific field at 1000 or 10.000 per week. With both figures it is possible to prove the contention that there is apparently sufficient opportunity for publishing and that one has a "free" choice.

But stop. Surely one should know that there is such a thing as censorship. The editors of any review that respects itself have an enormous task in reading, judging and after that accepting or refusing the articles sent them. If they take their task seriously, then they find themselves faced with an educative possibility and task which they can fulfill with the very greatest devotion. But there *is* censorship. A censorship that is fired with the very best intentions and that produces excellent results. For after all what would things get to if things came from the press indiscriminately. Nevertheless there is a threat of extrinsic restriction of freedom. Whether this occurs often or not is not important. In substance this restriction has been accepted. A dangerous symptom? It is said that there was once a prefect in Paris who said he would prefer a big scandal in his prefecture from time to time to having to bend his neck under the yoke of raiding and searching houses – quite a number of times in his district a year and at innumerable points! Society as we know it however, in its severe need through its ever increasing complexity, *had* to choose the latter way. Do not think, though, that the man whose conscience is clear, need not worry about it and is necessarily blameless. Why could it not be possible that he too should be brought into conflict and have to undergo a seach of his house? It is the same – in a figurative sense I mean – with the censorship on scientific publication. Imagine that it was not so. After all there *are* people with a lot of experience who are really able to sift the chaff from the wheat. Would it not be criminal waste to let the proffered oportunity to make use of such ability pass? That this includes the possibility of refusing an article on the grounds of scientific worthlessness but that later proved to be of first rate importance is all in the game. Naturally. It is all the work of humans.

But we have herewith again accepted an extrinsic restriction as a matter of course. Con amore even, I ask no more than that this should be recognized, so that we may be a little less surprised when we follow the investigation to the last stage, the stage that is so often forgotten:

What happens to a scientific investigation when the "finishing touch" has been given?

6. *The child of the brain has been given birth.* Care for this child of the brain is unknown to us, though. We are still at a very "primitive" stage. I should like to compare it to the egg of the salamander that is deposited in one folded leaf of one water-plant

in a ditch that is part of an endless network of ditches. Circumstances determine its fate. "Society" then has a free hand with the product.

In celebrating its bi-centennial Columbia University has raised the subject of "Man's Right to Knowledge and the Free Use Thereof". Our gathering resulted from this. The subject that now comes up for discussion to my mind forms the real theme of the above mentioned centenary celebrations.

We now have before us the data of the attained knowledge and we ask ourselves what freedom or restriction there will be when society has received the "free" disposal over what has been attained. Only here are we in medias res. I am convinced that all the foregoing did *not* come under the original intension. Nevertheless it was not possible to do without the extention that was here given to the subject, because when all is said and done there is a definite connection between the fate of a finding and the history of its origin. It also has an effect on the birth of the next finding, in other words on the idea. We have arrived at the critical point in the ever-recurring cycle, namely at its keystone.

It will be clear that problems do not arise in connection with insignificant findings. Or, to say it more bluntly, the only ones that count are the really important findings and the only thing that matters is what will happen to these after they have been made public. The mediocre findings disappear in the wake of the momentous ones.

The first question now is: shall science freely publish its findings unconcerned as to the consequences? Is it possible to prevent the man of science from publishing his system of thought and invention? Can one on the other hand force him to do so? We are faced with a horrible conflict arising out of a society that is definitely no longer able to keep up with the driving power that originated in itself, or rather that originated in some of the parts out of which society is composed. This society that has become so complicated is conscious of this and is oppressed by it. So it tries to establish rules in order to keep the system (i.e. the system of science) in check that contributed so much in causing the sense of oppression and it tries to make it subservient to its – not very clear – purpose.

Society, however, is at the same time obliged to cherish these servants because it can on no account do without their services, neither in maintaining its way of life, nor even in preserving its continuance.

Science likes being cherished. High up in the ivory tower it is cold and lonely, one can hear science calling to itself that it may not stay put there. And thus, of its own free will even, it has extended its aim for serving truth only into serving society as well

Society is afraid of this servant, however, which it can no longer let go.

One may now hear it argued in many different keys that technics is the main "culprit" in this dilemma. In all modesty but also with positiveness I should like to stress that medicine has had a very great share in this. I tried to elucidate this more clearly on a former occasion [1]. The branches of science that have man as their object have their own goal in view from the start, namely to put at man's disposal all they are able to produce in the way of help. Society, afflicted alas with continual misgivings about itself, cannot desist from snatching away from medical science as it were, the real or fancied results it has attained to. Medicine itself is more reserved, at least in its best moments. I would like to pass by the moments of weakness in which the research worker, who believes he has made a find, publishes it before any proof has been furnished, simply because it gives such a beneficent feeling to have found something. And the weakness of medicine itself, which in its ardent longing to help, remains unshaken in its credulity. Fortunately there *is* still such a thing as healthy scepticism, which keeps in check the too frequent opportunities for derailment.

And thus, when the question is put whether one should freely publish, the difficulty arises, especially in medicine, whether one should not be very reserved, because the certainty that has been attained is still insufficient. On the other hand whether one should not quickly make public what has been attained, because there is the chance that one may be able to help many people. All right, it is still possible to accept this restriction in freedom completely, on account of its large intrinsic component.

But it is different when pressure is brought to bear on the

---

[1] In "Wetenschap en Maatschappij", (Science and Society). Holl. Maatsch. d. Wetenschappen, Haarlem, 1952.

research worker. Examples of this exist both for good and for evil. When Koch had let his studies on tuberculosis go in the direction of immunisation, tuberculine appeared as the product of his efforts. The idea of being able to cure the disease, which at that time was very much more disquieting than it is considered nowadays, was expressed and it seemed that there were results. Koch, however, remained reserved. But the people in his environment appealed, curiously enough, to that same feeling that caused him to be reserved, namely to his sense of responsibility towards mankind. The result is well-known. Koch yielded to the pressure and it only hastened the death of innumerable patients. This was a dramatic but short period because the results were only too clear. There are other examples. Did not the introduction of X-rays into medicine, by the side of its many possibilities, cost numberless patients their insidiously injured lives? Medicine repeatedly has to start publishing with insufficient knowledge. Who shall make out to which side the bondage to both aspects of the self-same duty shall turn the scales?

From this comes the question: who is responsible for the decision, however it may turn out? Who is responsible for the consequences? Here everyday human weakness comes into play: if everything goes well it is readily accepted; but if things miscarry, a great outcry is made and everyone wishes to make someone responsible. It is the same with respect to science. The utterances are numberless that make it responsible for the disasters that it has actually, or potentially brought upon mankind. The atom bomb is the most topical and most navrant example of this.

In so far as it concerns medicine, it is the bacteriological war. There the problem of the freedom of Medical Science comes acutely into play. On exactly the same basis as that on which aid was offered to suffering humanity that it might be liberated from the scourge of murderous epidemics, on that same basis the means are thought out which, while protecting the home camp will, if not massacre that of the opponent, at the very least demoralize it. Here there is no question any more of any freedom being left over. The (medical) bacteriologist is caught in the web of the inescapable. If he does nothing then he will fail those people who will presently be victims. He will fail his government and his country, which shall presently ask for his help when the develish menace shall be let loose. If he puts forward all his abilities with the intention of collecting protective forces, then it is quite inevitable that he will have to reconnoitre in the field of the agressive forces as well. That makes his impasse complete. If he wishes to help then he helps to bring the evil to pass, if he refrains altogether then he helps – indirectly – to give free rein to destruction. His protégé is his victim and he himself is more or less manifestly under a charge: the prisoner of his protégé. One can turn things this way and that but there is no way out for medical science. There is even a more absurd consequence still: if the work of bacteriological warfare is taken in hand, then it is by no means beyond possibility, or rather it is probable even, that fresh possibilities will be found for helping suffering humanity without any thought of warfare, just in the same way as destructive possibilities arose out of the primary intention of helping mankind. Must not we, and thereby I mean medicine, contend that we owe a great deal to the wars: chemotherapy with the antibiotica of Penicilline discovered during the second world war and aureomycine discovered after it? Have not fresh prospects come – modest still – of approaching cancer through the means of fefence against mustard gas (the "B.A.L."). Have we not seen research possibilities and chances of cure arise out of the splitting of the atom as should be considered unrealizable outside of a war. The contention that psychiatry has progressed by leaps and bounds to a point that would otherwise not have been attained for decades to come, I should like to place in this connection – without being able to vouch for its truth, however –. It was also the war that gave us D.D.T., together with a whole series of new drugs and new knowledge about the causer, which has enabled us to establish a great gain in our fight against malaria. Was not the opportunity offered us in the war of studying many new things in an "experiment on people" to an extent never displayed before? It is indeed the most conflicting position that medicine has ever experienced: it owes an unutterable amount to the help it *had* to render in a situation that filled it with the most elementary feeling of loathing. The fulfilment of its duty as medicine allowed medical science to benefit from morally unacceptable possibilities.

One cannot even confront science-for-its-own-sake with such a situation.

As soon as the results of science reach society there is both joy and a feeling of oppression. Medicine has prolonged thousands, nay millions of lives and filled them with joy, but together with all the other sciences it has also helped create the problem of age, of competition, of the division of productivity, of the disturbance of the economic equilibrium, of the food distribution, of the limitation of energy, of over-population in short of the *crowded state* in the world.

Out of this confusion the student in the medical sciences will have to find his own way, for the simple reason that he *must* go on, because he is the student and the object at the same time. One cannot bring medicine to a standstill and leave man's suffering in times of war to run its course. Yes, for friend and foe should be equal ... but now I enter upon the field of medical ethics; I shall leave motives out of consideration.

Even if one were to look upon that as a way out, it would yet be *impossible* to put a stop to the progress of medicine. We cannot choose, we *must* continue. One may object that this holds good for every branch of science. I do not believe that that is so. For have I not heard secret approval accorded to a German scholar who could have made the atom bomb during the war and put it at the disposal of the absolutist régime, had it not been that his moral consciousness could not brook it. So here development can be retarded, if only temporarily. A certain purpose *can* be served by postponement. In medicine that is impossible, because there one is faced with an inner conflict. Where can we find an indication as to how to act?

No longer in objective thinking, in rationalizing about science, about its practice, its aim, its duty, its right, its freedom and restriction.

The way to act can only be determined by a sense of responsibility. And this we can only reason out from the standpoint of our own ego. It is only the individual who can try to merge it in a higher aim. A multitude, society simply has no sense of responsibility. For it can change with the force of the arguments that are poured out over society and which give it this or that shape. Even when an individual confronts his own interpretation of things on the strength of his sense of responsibility, with the way of thinking of one other single individual it may appear that there is a radical difference in insight. What then can we expect of the broad masses? The greater a multitude is, the smaller will be the change of a point of contact. The greatest common devisor of a big number of figures becomes smaller and smaller until it shrinks to the unit which, as has been agreed upon, no longer counts as such.

As long as science therefore directs itself towards the broad masses it will find neither freedom nor support for a sense of responsibility. So long as science knows that it is subservient, it is bound; so long as it wishes to be useful it cannot be otherwise. That need not be a disadvantage at all, if its student has chosen this to be so. Then he can determine and find his circle of freedom within the restriction which *he* can experience as intrinsic. Along the boundaries of that circle there are stakes all of the same pattern: the feeling of responsibility. These stakes are only visible to the student himself, everyone else walks by those stakes unseeingly, in his turn seeing only his own boundary posts that have been raised by his own feeling of responsibility. It is not pleasant to know that there is something so ungraspable in one's way. There is nothing else for it but to experience that since our standards have ceased to exist, life has only become more difficult and curiously enough has not broadened our *feeling* of freedom. We should so much like to pass *our* own stake on the authority of *another*. When only our *own* boundary posts surround us we feel lonely.

Perhaps it is for this reason that we so much like passing by the boundaries of another on our own authority.

Will it be possible to succeed in getting science, that is to say all its students or at least all those of any importance among them to assume a certain unity as to shape and dimensions in the feeling of responsibility? Even if it were only among the members of one team? The church has been obliged to give in and science has not offered any substitute, not even the science that serves man's welfare most closely. We do not know. We only know this: that there is at least one road-sign pointing to the goal we are striving after; I mean absolute sincerity and disinterestedness. With its help science will perhaps be able to determine its responsibility and act accordingly. Will this be possible? Will science, which also tries to fathom man's driving motives, be able to trace the causes of insincerity and vanity, to recognize them as pathological and bring about their cure? Or will the road of short-circuiting be the only possibility? This would mean: again accepting the dogma, whatever its content, to attain unity in the feeling of responsibility.

Until such a time the student of science will only be able to experience a feeling of freedom when he sees the spark of his idea flare up, and when he is facing his object. Those are his best moments. Once the child of his brain has been born, it will lie shackled by society and he himself will know, that he is fettered too. Only the – solitary – following of his own feeling of responsibility will give him freedom, even if he is called to account. Whether we like it or not, that is how things are. As a medical man he has over and above this the favour of serving: a chance the more of giving him the *feeling* of freedom in the pursuance of his branch of knowledge.

DEBATE ABOUT PROF. H. W. JULIUS' PRE-ADVICE ON "THE CONSEQUENCES OF FREEDOM AND RESTRICTION IN MEDICAL SCIENCE"

by M. C. COLENBRANDER

The words I am about to speak will not be words of criticism but of approbation. Approbation of the great importance *Julius* attaches to the *personal* element in science, an importance that makes him rank the man of science with the artist.

I hope I may be permitted to pursue this facet further. Beforehand I feel obliged, however, to bear testimony to what to me is essential in science.

For it is apparent that science is variously judged, also among its own students, both as regards its intrinsic truth, as well as its moral task. This is also demonstrated in these pre-advices. *Koningsberger* is convinced that absolute truth does not exist. He says: "In reality the natural sciences are no more than the work of *homo ludens*, a playing of the human mind which is only turned into an art by the very few".

Pos on the contrary says: "What is fundamental here, is that it has proved possible for the human mind to arrive at objective knowledge". He who wishes to stress the personal element in science incurs being suspected of denying objective value to science. I believe mistakenly so. In science absolute and relative are inextricably mixed.

I should like to explain more fully. To me geometry still counts as the prototype of science. Starting from premises it arrives at compelling conclusions through logical reasoning. The premises go back to unprovable axioms. I believe this method can be discovered in all science. It gives facts and it works on those facts with its instrument, logic. In so far as there is uncertainty in science it is due to the premises. The laws of logic are unassailable.

All science strives after objectivity. It lays claim to a general validity of its opinions and also of its doubts. This is thanks to its instrument, logic.

This logical thinking is necessary for drawing conclusions from observed facts, but also for thinking out experiments, so both down stream as well as up stream the river of causality.

Now to my mind logical thinking is much more difficult than many people suspect. For our thinking is by nature guided by associations of a personal character, amongst which our appraisements make up an important part. If a possible conclusion clashes with an appraisement then there is a great chance that that conclusion will not occur to us. That which has an old familiar ring is positively accentuated in our appraisement. And for this reason it may happen, as *Wertheim* writes: that "man has a feeling of resistance to all that is new" and that "Practically every scientific renovation is the result of a fierce struggle against formal knowledge" One should not forget that his deficiency saves civilisation from too great shocks. Without powerful customs there can be no strong culture.

Often too a newly found relation leads to an over-estimated idea. How frequently does one not forget that a newly discovered causal relation does not imply that the discovered cause is the only cause of the phenomenon in question. The discovered cause all too soon receives a monopolistic place. The conclusions that are then expressed, are marked by such words as "only" or "nothing but". The less one knows of a subject the sooner one allows oneself to be tempted to utter such apodictic words. Especially in psychology, a field about which experts know little and laymen nothing at all, is a field where such attitudes are rife.

The task of scientific thinking is to wash out of the net of associations all the personal fibres, so that the logical connections which are impersonal, but which for that reason are of general validity, can come to light. That is the task of every scientific thinker and this does happen at the universities.

The logical network – to stick to our image – is of more than one dimension. But, argument only has one dimension. In the manner in which the demonstrator proceeds from one knot to the next, going through the whole net, may lie a great artistic and personal element. It is in this respect that the man of science may be an inspired artist.

One more word I should like to devote to freedom. Thinking and feeling cannot be robbed of their freedom. The problem of freedom only counts for one's actions.

All actions are the result of motives, and very often more than one motive play their part.

The motives are weighed against each other according to their importance. The image of the scales is here entirely in place. When there is talk of freedom of action we mean that some motives lose their validity as determinants of that action. The other motives automatically receive relatively greater importance. If the scales are tipped there must be a weight somewhere.

Freedom of action is usually understood to mean absence of restraint from without, so that the action is entirely determined by one's own inner nature.

Now one's inner nature is not an indivisible unity. One may be in conflict with oneself, for instance when the urge towards knowledge and reverence for life influence us in contradictory ways. When our own moral standards or when matter itself form an impediment to our urge towards knowledge *Julius* calls it an intrinsic impediment and he rightly considers it acceptable. Yes, we cannot do otherwise than accept, because it is impossible to undo us of ourselves. We can withdraw ourselves from the moral standards prevalent in our environment if we wish, even if they feed our own weak moral standards. But then the standards have become extrinsic, from without so to speak. The problem of freedom and restriction is only a question really of the acceptability of the extrinsic ties. In the beginning of his argument *Julius* says: "We reject extrinsic restriction". Fortunately he retracts part of this further on.

To my mind the classification into intrinsic and extrinsic restriction is not very illuminating, because the boundary line between what is and what is not permissible does not lie between these too, but it lies right in the middle of the extrinsic restrictions.

In the pre-advices a great deal has been said about the relationship between science and ethics.

Pos says that: "the task of the research worker is an intellectual and at the same time a moral one".

Julius wonders whether science should freely publish its finds, without thought for the consequences. He sees here – and so do I – a hideous conflict.

Van Melsen on the contrary says: that "the student of science is not responsible for the abuse to which his knowledge may be put. His primary and most important responsibility must be looked for in his striving after truth. Every other responsibility resting on him, he has in common with other people", whilst *Langemeijer* is of opinion that "where real, genuine science is concerned, aware of its own limitations, there can be no question of its coming into conflict with genuine morality".

Morality is unthinkable without freedom of the will. This brings us to an extremely difficult field. *Julius*, very rightly, is cautious and only speaks of the need of a sense of freedom,leaving it an open question whether this feeling corresponds with something real or not.

I have given shape to my own view in the following image. An image often shows the point better than a long argument would do. An image tries to express how the impossible may be possible all the same.

A traveller from antiquity wishes to go to the ends of the earth, to which end he travels North. When the region through which he journeys becomes uninhabitable he now and then puts a stick into the earth so that when he looks back he can convince himself that he has deviated neither to the right nor the left. At last he again comes to habitable regions and inquires after the direction in order that he may be able to correct possible deviations. To his utmost astonishment he hears that he is going South.

For us, who know that the earth is round, the solution is not difficult. But this man from antiquity, who in imagining the earth, had to make with one dimension less, must have wondered in vain, how it is possible in starting from a well-chosen point and walking straight ahead, to miscalculate so, and how it is possible that North and South, which are every where contraries, unexpectedly come to be each others extension. This image shows that misunderstandable riddles become understandable for the person whom it has been granted to think in one dimension more. When applied to the problem of the irreconcilability of causality on the one hand, which says: "everything is determined, there is no choice", and responsibility on the other hand, which says: "You *must* choose", it shows us that this contrariety might perhaps be capable of solution if our powers of thought had as it were one degree more of freedom. But it is not so. Our power of thinking is limited and the limitation consists in the fact that our thinking in terms of causality is a necessity; and every necessity is a restraint and every restraint a limitation.

This limitation is otherwise not without analogy. Nowadays we know that there are physical phenomena which can only be understood by the assumption of curved space. Such a curved space is not imaginable because our imagination only has three dimensions. It is my opinion that science which rightly postulates the unlimitedness of causality, and the sense of responsinility, that rightly postulates the necessity of free will, are engaged in an unsolvable conflict, and I look upon the unsolvableness of this conflict as the result of our limited power of thinking.

As regards the practice of freedom and restriction in medical science I can only whole-heartedly agree with what *Julius*, says in his pre-advice. Reverence for life prevails over our urge after knowledge, even if on account of the gratification of this urge the lives of others might in the future be saved.

So it is of set purpose that we impose restraints upon medical science. But should one expect otherwise of people who have chosen as their task in life to be protectors of life?

DEBATE ABOUT PROF. H. W. JULIUS' PRE-ADVICE ON "THE CONSEQUENCES OF FREEDOM AND RESTRICTION IN MEDICAL SCIENCE"

by B. H. KAZEMIER

A perusal of Prof. Julius' pre-advice has strengthened me in the conviction that, whilst it may seem as if the problems under discussion at this congress have become of current interest for the students of the various other sciences only under the pressure of the social and political circumstances under which we live, these problems have really always been well-known to the students of medicine. The explanation for this is not far to seek: medical science, however much it strives after truth as do the other sciences and however much it is directed towards the attainment of knowledge, has yet by virtue of its ultimate objective never lost touch with the reality in which we live and die. It is in this reality, with its care for our daily existence and fear of death in which it is not abstract knowledge that comes first but action, that we meet the physician. In this concrete reality of action too, the idea of responsibility first becomes important; for Man is responsible for his actions. That the medical student has always been aware of this responsibility appears from the fact that since the time of Hippocrates there have always been medical ethics by the side of medical science. In the pre-advice the problem of freedom and restriction in this double aspect of scientific truth and ethical responsibility has been raised. Regarding restriction in the pursuit of science the speaker has distinguished between two sorts of restraint of freedom which he has called intrinsic and extrinsic. The two forms of restriction that the speaker has thus voiced, and his warning against the confusion of ideas which might result if one were wrongly to identify the two sorts of limitation of freedom, I should like to endorse, even though one might well doubt whether the historically overtaxed words intrinsic and extrinsic are the most suitable terms for expressing this distinction without misunderstanding. But the matter itself is more important than this terminological question. And here, I can see, the speaker has done a good thing by warning against the confusion that does indeed often occur between scientific restriction which has its roots in reality being thus and not otherwise, and moral restriction which we experience as an inner obligation. When we try to reduce these two sorts of restriction to the same denominator we misjudge the difference of principle between fact and standard. The serious consequences that may be the result of not considering this distinction have been strikingly illustrated by Prof. Julius with some examples taken from the field of his own branch of science. By so doing he has also placed the students of other branches of science under great obligations. For this same confusion of ideas is also to be found in other sciences. For instance to give one example, there is the confusion between the science of economics and political economics, a confusion of ideas of which both the supporters of economic liberalism as well as orthodox marxists have been guilty.

That the problem of freedom and restriction in science is still far from being solved by a methodological distinction having been made between scientific determination and moral responsibility is eloquently but also painfully evinced in the speaker's argumentation. Prof. Julius as a matter of fact has modestly qualified his pre-advice as an attempt at a definition of ideas, even if it be not only for medicine but also for its object and its exponent, Man.

In treating his subject Prof. Julius has put it thus that freedom in science is experienced when the individual thinker is occupied with the pursuit of science in its narrower sense; when he is in the act of conceiving new ideas and checks his hypotheses with reality. The restriction is not experienced by the scientific investigator until afterwards. Once the child of his brain has been born, so says the speaker it is chained by the bonds of society and he knows that he himself too is shackled. And that is how it is. Which of us has not undergone the fascination that a scientific problem gives and the liberating effect its solution has? Objective truth is free from human arbitrariness and the subject has risen above its individual limitations in so far as it shares in this truth. On the other hand, how could a scholar do otherwise than consider himself entrapped by society when the material means are withheld from him with which to pursue his scientific research to the end, or when social forces try to drive scientific development in a particular direction? How could he feel free in a society in which scientific truths are not accepted because they do not tally with traditions generally upheld or with ideologies that are looked upon as sacrosanct? And then a question of greatest current interest, what freedom remains to a scientific research worker in a society which aims at using the results of his knowledge for purposes of wholesale destruction?

And yet this view of society as the cheater of freedom in science and of its students, as we meet with it in Prof. Julius' pre-advice, is not free from one-sidedness. In order to arrive at a fair and perhaps less somber appraisement of the significance of society for science one may not lose sight, so it seems to me, of another truth, namely that it is human society, that makes science and its practice possible. Once there is an organized community of sufficient stability, in which a certain level of prosperity has been reached and in which a certain division of work is possible, the conditions have been created wherein science can come into being and be maintained. This naturally also holds good for medicine. And therefore, however much society may be a menace to the freedom of science, it is no less true that it is society that makes the free pursuit of science possible in the first instance.

Nevertheless it is obvious that it is the menace to his freedom that in the first place will be the occasion for the intellectual to reflect on his place in the social system, in the same way that we are usually not interested in the functions of the body until we are afraid for some possible disease. And how is the research worker to arm himself against this menace? It is here that Prof. Julius appeals to the sense of responsibility of the individual, whilst explicitly rejecting every attempt at finding one's bearings by the broad masses of the people, or by society because neither the one nor the other has any sense of responsibility. A restriction which the scientific research worker may experience as intrinsic, the speaker considers entirely individually determined. Everyone puts his boundary-posts in hiw own way and does not acknowledge the boundaries someone else has drawn on the plea of *his* sense of responsibility.

It seems to me that on this point too Prof. Julius' view is one-sided. One-sided, not incorrect. Because it is indeed true that no one can shift the responsibility for *his* decisions and *his* deeds on to the shoulders of another, not even if he does not let himself be guided by his own insight but acts on the authority of a spiritual or worldly leader. Because then too he remains responsible for accepting as binding some religious dogma or political ideology. And yet I should like to ask whether the feeling of loneliness that may seize us in the taking of decisions, especially of those that cost a great effort, are necessarily as great as has been made out by the speaker, or so absolute as existentialism would have us believe. For do we not remain part of the community to which we belong, even in the making of a choice which we feel to be strictly personal; and the convictions held in that community and the standards generally accepted there will influence even our most individual decisions. Nor do we by any means always experience this influence as a resistant. We may also feel upheld by the realisation that the standards accepted by us as binding, have their roots in the reality of which we form a part. And on the other hand, it is true that they whom we honour as the great pioneers of mankind in scientific and other fields, often had to fight ideas which in their day were generally considered true and against standards that were generally accepted as binding – Prof. Julius reminded us as a striking illustration of this of the prohibition that was valid four centuries ago against post-mortems – but their novel views, at first put forward at their own risk, and the new standards applied on their own responsibility, have later been recognized and accepted by a grateful society.

The problem of freedom and restriction in medical science has been described by Prof. Julius as the problem of relation between the individual scientific researcher and the community. This relationship has become more problematic probably than ever before, on the one hand as a result of the unsuspectedly rapid progress of science and technics, on the other hand as a result of the social and political tension in present human society. And yet there is perhaps this aspect to give us hope, namely that this problem is consciously experienced as such in scientific circles, not only, see this congress, in our country, but elsewhere also. Because even though the questions that have come up for discussion with the theme "Freedom and Restriction in Science" are not, if I see rightly, themselves for the most part scientific problems, but rather ethical and political ones, this does not alter the fact that here too it holds that if a problem is rightly put it is the first step towards a possible solution.

RESPONSIBILITY IN THE HUMANISTIC SCIENCES

by H. J. Pos

The classification of the sciences under those belonging to nature and those pertaining to the mind or culture is justified in fact. It's practicability is continually being proved in research and education. The fact, that there are regions that lie on the border of both these fields only goes to show that nature and culture are distinct but not opposed to each other, nor that they are without any inter-connection. Man sprang from animated nature and everything archieved by mankind bears traces of that origin. Biologically mankind is permanently bound to nature, even though it is developing more and more into a reality of its own.

The humanistic or cultural sciences investigate this reality in its past and present state. History is one of them, extending from prehistoric times to the present day, and also embracing social relations, technics power, law, morals, property, language, art, religion. These realities appear in two shapes, as living, ever-changing forces and as objects of research. Between these two there is a tie and also tension, which latter is caused by the fact that all branches of knowledge are themselves a part of human life. At the point where these two figures meet, the researcher of the cultural sciences finds his task and his responsibility. Seen from a historical point of view this meeting first occurred in Greek civilisation. It became typical for European civilisation, whence, during the last centuries and especially in ours, it spread over the whole of mankind. What is fundamental here, is that it has proved possible for the human mind to arrive at objective knowledge. The result of this fact is that in all the spheres of reality knowledge strives to arrive at an image of things that is objectively justified. There is another image that precedes this one and that is able to vindicate itself to a certain degree, whose function it is to serve life. Ideas about the past play a large rôle in the lives of families, groups and peoples. In this way traditions are handed down from the older generation to the younger generation. The company of other children together with education arouses images that are at the same time dispositions for behaviour, like for instance the prejudices regarding others and strangers in the widest sense of the word. The emotional attitudes thus aroused, usually rest on opinions about people and about past events. They seem to borrow their being from the ease and generality, with which they are handed down and passed on. The traditional collective image has an effect whose power is involuntarily taken as proof of it's truth. According as its power is confirmed in age-old traditions the need is felt less for questioning the objective soundness of those ideas, which seem to demonstrate their truth in life by the impression they make and by the services they render.

To determine the degree of the objective soundness of traditional ideas falls to the task of the research worker in the cultural sciences. He must be on his guard against the abuse that is made in life of the past by individuals and groups. The objective determination, it is true, is the business in the first place of scholars amongst themselves. There is an internal responsibility which places the practitioners of the same branch of science under mutual obligations and which constitutes a mutual tie. But a higher and more all-embracing task of theirs is that they as scholars shall place truth over against the image of the past that is determined by the life impulse and by inborn egocentricity, in so far as this is attainable with a reasonable amount of effort. The factors that may be counted as belonging to the pre-scientific image are: the striving after power and self-assertion, the wish to whitewash, cleanse and conceal, and also to exaggerate and caricature. And then there is also the striving to worship and serve, and to find one's own ideal in others or to project it into them.

History lies at the heart of the humanistic sciences. It presents a development in phases, none of which have disappeared for good; besides occurring as a succession as of increasing qualities these phases also occur simultaneously and without any special order. In our own century for instance we have things come back that seemed to have been conquered for good. The origin of historical consciousness can hardly be thought of otherwise than as a store-house of events, that made a deep impression, for instance heroic deeds or calamities. Tribal chiefs worshipped their ancestors and by keeping their memory green, they perpetuated the halo, whose splendour also reflected on themselves. Homeric heroes and the great among the Romans traced their pedigree to the gods, and the Emperor of Japan is the so-manieth descendant of the Sun. Marriages between gods and men repeatedly occur in pre-historic times, as also the idea that the gods are greatly interested in the fortunes of a ruling family, tribe or nation. Jahveh brings the Israëlites out of Egyptian bondage. He divides the sea for them and helps them conquer a country that is inhabited by people who do not honour him. He proves his superiority over Baal when he lets fire rain down.

Such subjective interpretations of historical facts are called mythical, i.e. for a more objective opinion the reality of such interpretations are confined to their effect on the lives of those who accept them and they are identical with it. This does not by any means imply that such interpretations no longer occur. Twenty years ago the myth was rampant right in the middle Europe that a race felt chosen to dominate Europe. Another point on their program was the annihilation of another race to whose account they put down all the preceding disasters. It is true that in this mythology the accent was more worldly and the disposition of its exponents was less disguised. What was essential in both these cases and what they had in common was that the urge to power with pure hate as its companion, together thought out the ideas suitable to nourish this urge, found arguments for them and circulated them.

Whenever he sees the natural life impulse, that is bound to limited aims, at work, there the professional histiorian has his task. He should keep on asking himself whether the idealized image of the past, that he has about family, class, nation and race is covered by the facts in so far as they can be verified. He must also keep on recalling that truth is much to seek in the motley mixture of leaders, groups and institutions, one and all convinced that they know the absolute and irreplacable meaning of things. He should treat the contradictory claims, which are not concerned with each other, with scepticism and he should require them to be heard and judged before the bar of reason. Such an examination never recognizes the claims of the contestants, nor of even one of them. In so far as it concerns the ideas that have been raised, reason never expresses itself otherwise than for a n d against. It distinguishes the truth from the exaggeration and changes the negative qualities of absolutism and exclusivism into the positive qualities of relative tenability and gradation. It does this equally for the past and the present. The active and emotional attitudes of men and communities in their relations to one another have a tendency to accept or reject an image absolutely because life even if it is not best served so, is yet the most easily served to. But primitive decisions are coupled with a coarsening, respecting truth. A striking characterisation of what goes on in the mind of an educated citizen of a nation that is involved in a war with another nation of whose culture the man in question is a connoisseur, is given by Bergson in "Les deux sources de la morale et de la religion" (1939, 28th edition, p. 309): celui qui connaît à fond la langue et la littérature d'un peuple ne peut pas être tout à fait son ennemi. Experience has taught us that the admiration for what a given nation has produced in the field of art and learning, sometimes prevents its connoisseurs from seeing the full objectionableness of the deeds of aggression and terrorism directed against their own country.

From the time of Greek Antiquity onwards a process of critical awareness has been in progress in our civilisation regarding the standards to which rational judgment subjects beloved traditions. One cannot expect a critical sense from epic writers such as Homer, nor from the oldest historians. They were wont to recite at the courts of kings, who liked listening to the glorification of their ancestors. The Iliad and the Odyssey may have a historical nucleus, nevertheless the fictitious element knows no bounds, especially in the latter. Hellenic historiography as a whole did not rise above the useful and the aesthetically appealing. Even the great Herodotus is unable to choose between traditions that contradicted each other, the one of which is even more improbable than the next. But the critical and realistic Thucydides rises head and shoulders above his colleagues. It was to take till the 18th century before the scientific method of writing history that had been put into practice by this giant in his trade, came to be the recognized one. Not until the Enlightenment was a clean sweep made of myths and legends, after a few solitary precursors in the sixteenth and seventeenth centuries had expressed a critical opinion on witches and demons, magic and miracles, and then only very tentatively and with danger to their lives. In that period the Copernican turn, as Kant called it, took place: it is not things that make their existence felt on readily receptive minds; not every report, however extatic and sensational, is worthy of credence, for there is a standard that determines whether a tradition c a n be true or not. This standard is the best of reason, of which Descartes taught that it works in every one of us. A famous poet c a n n o t have been born in all the seven towns that claim that honour; a historical personage c a n n o t have lived in a certain age as well as a hundred years later; a battle c a n n o t have been won by both sides; a human individual always has human parents; an eclipse of the sun or a comet have no connection with happenings on earth. Reason certainly does not determine the *content* of history, because to do so experience is requisite, but it does determine the formal possibilities, two contradictory things cannot be true at one and the same time and in the same respect. All this might already have been found among the Sophists and in their great pupil Thucydides, but it was now discovered afresh, and this time it was to become part and parcel of the general scientific awareness. In that century of emancipation (Kant) it was to demonstrate its full fecundity for good. For good however, was to prove to have been saying too much. For in the first place rationalism was to be followed by romanticism, which was to confer a fresh place and fresh values on what had gone before the critical period, that period which had been pushed aside by criticism. Romanticism thus involved truth itself in a process, thereby weakening the distinction between truth and opinion. If truth is not the bar before which all special views have to appear, but if truth is their process, who will then still be able to distinguish truth from mere views with any degree of sharpness? This is how romantic dialecticism put the problem. If one draws this line further, then everything is equally "true", there is no longer any criticism, and no effort is necessary, every opinion has its own right to exist. With this night has come, but everyone has a light of his own making and that is enough.

The charge that it kills feeling, which is a basic value of historical thought, is often brought against the rational way of writing history. It was this that Romanticism protested against. In 1939 J. M. Romein put this problem somewhat differently in his talk on "Het vergruisde Beeld" (The shattered Image). To my mind the honour one shows irrationalism, is too great if on this point one only bemoans a loss. A higher truth can never be sad and a more restricted truth can never make entirely happy, even if it fascinates. With the transition from more restricted and more fervent into a cooler and more all-embracing atmosphere we find ourselves in a process involving the whole of mankind. The transition from the writing of history from a romantically biassed view to its writing in a critical and universal light is a symptom of this process: the driving motive lies in the striving of present-day humanity itself to attain unity. We, who were still brought up in the doctrine of the sovereignty of the state, were given an intensification of collective particularism in that doctrine. We can no longer speak of the truth of this doctrine, for the actions of so-called sovereign states are far from independant now-a-days. Probably they were no longer so in our case either, while a chance configuration in the balance of power increased the semblance. In any case the fact has got to be accepted and we must try to make the best of it. An advantage in this respect is that the vision on the powers determining the course of history is certainly widened by the factual refutation of the delusion of independence and safety, which we witnessed. Historiography and teaching can only profit when they realize that the state is not a monad or a substance, nor independent, self-determining and inaccessible, but that it is open and reacts sensitively to the whole of which it is only a member; longing for contact with others in exactly the same way as a human individual is not a self-contained unit, but rather a nodal point in a limitless web. The research worker in the "Geisteswissenschaften" is responsible for the diffusion of knowledge in a spirit which not only does not stand in awe of chauvinism, imperialism, racialism and all kinds of fanaticism, so of limited things that have been rendered absolute, but which roundly denounces and attacks them.

A sort of depressed feeling sometimes besets present-day scholars of the humanistic sciences. It consists in their feeling that the research has become futile, which has been freed from its former tie to limited aims like class or nation. They would like to return to the earlier way of doing things, realizing at the same time that such a return is incompatible with the truth that has meanwhile been attained. This feeling of sorrow would disappear, if the realisation were to be reinforced that the task of the research worker is an intellectual and at the same time a moral one; that he is called upon to serve truth in the clash of interests and their accompanying prejudices and delusions; and that he should not wait to see whether he is consulted or not, but that he should let himself be heard of his own accord wherever passion menaces free judgement. If he is alive to this realisation, then the boredom to which he feels condemned as an onlooker and recorder of truth, after he has attained some academic function owing to his ability and industry, then this boredom will change into an activity that will overwhelm him and that may even be too much for him, but which at any rate blesses him with a pregnant task.

One aspect of how it should not be done in this respect is still alive in our memories from the years 1933 to 1945. Teutonic irrationalism had already long been hammering on the tenet of the subjectivity of all moral principles, on the accidental character and the equal rights of all action: in this way the most revolting expropriation and criminality could show its face with a good conscience. This irrationalism was indirectly supported by some brand of religiosity, which had always taught that man is at bottom develish, and which now seemed to be confirmed in its creed and hitherto withheld by experience. The revision of our Dutch historybooks belonged to this period. They now started to teach that Holland, which had been torn away from a larger empire in 1648, had undergone three centuries of decline, but now seemed to be on the threshold of a new beginning. The only unwittingly true part amidst all this falsification was the formal claim to see the history of one's own country within a larger framework. However, this requirement had already become conscious in the minds of others as well. During the occupation many books and papers were published, which evidently had the object in view of comforting the Dutch by reminding them of their former greatness and the free culture peculiar to their country. That the occupiers permitted this was probably not due to their magnanimity: the r e a d e r s fell into dreaming about the past, this sort of reading had no special effect that might become dangerous. The state of the country after the occupation goes by the name of recovery, and this has certainly in many respects been so, in all its ambivalence. But just as this appeared changed in the way of v i e w i n g history, causing national history to be considered not only from within, but also from without, so from the point of view of what it looked like within the framework of Europe and mankind, in that same way something had changed in the a t t i t u d e because European countries were striving after a federative European relationship in which they hope to continue the forms of life which they know will be threatened as long as each country tries only to maintain them for itself. The connection between this striving and a wider view on national history is evident: now that under the power of circumstance the preservation of sovereignity cannot possible be wished for any longer because it is unreal, so also has the insight become freer in respect to the mistaken national idea of sovereignty. Now that a united Europe is being w i s h e d for, it is also being s e e n as a reality, which it already was really but which only required conscious consolidation. When one looks back from this state of affairs on what Nazi Germany professed but could not and would not bring about: a united Europe, the domination of 1940–45 looks, after the event, like a caricature and abortion of something great and good for whose realization qualities are required that were entirely lacking in the then leaders.

One may ask whether these observations do not fall outside the field that is defined by the concept responsibility in the arts and sciences apertaining to civilisation. I think not, in the first place because responsibility is a moral category which is solely applicable to living human beings. Our subject therefore must be understood that it is not "the arts and sciences" which may be called free or restricted or responsible, but only its practitioners. In an internal sense they are responsible towards each other as colleagues who do not evade criticism, but who ask for it. Their responsibility towards the outside world is much more comprehensive, towards all the use that is made of knowledge and learning and towards all the appeals made in that quarter. This use has various forms which coincide with the spheres of influence which the individual undergoes from his birth onwards: opinions and emotional attitudes in the home, in the intercourse at school, in the spiritual and material relationship of church or class in which he finds himself placed by fate. The influence of milieu and education during the susceptible age is well-known. Whilst the factual knowledge that has been imported is mostly soon forgotten, the moral influence of the educators and teachers on the other hand remain. In Holland freedom in teaching is great, and it implies that children to a very large degree receive tuition in the spirit, whether denominational or otherwise that their parents wish. This of course does not mean that they do without generally valid and suitable knowledge, but this is not aimed at in the first place. It is much more merely a means whereby each group strives to prove that their results are so good that the child is safest in their hands. This striving unconsciously justifies what it consciously denies: the meaning of rational objectivity. In the meantime it is unavoidable that this private basis plays its part in the treatment of certain subjects that cannot so easily be treated apart from one's appreciation of life. Teaching of national history in Holland offers examples of this in plenty. Protestants stress the religious quarrels as being the quintessence of the secession from Spain, and they are inclined to represent the embracing of the Reformation by individuals and communities as a deed of conscience. Catholic historians on the other hand stress the fact that in those days to become a Protestant was as much a business of collective coercion and in people's own proper interests as it was to remain a Catholic. They also call attention to the fact that religious persecution and oppression cannot be exclusively laid at the door of the mother-church, since she also came in on the other side at the changing of the distribution of power. The struggle in this field has brought with it that through the united rejection of neutrality, objectivity as a value and an ideal has also suffered and that feeling for it has been weakened in the education of the mind. In the years of Dutch political neutrality of before 1940, this struggle took such possession of people's minds in Holland, that it was possible for them to believe that it had its parallel everywhere among mankind, and as if there were a dividing line between belief and Right-wing politics on the one hand, and unbelief and Left-wing politics on the other, with Holland in a manner of speaking as a model example of this. Now it is true that in the present-day world there is a deep gulf and it is indeed possible to subordinate smaller contrasts to one or more larger ones. This predominating discard is also little covered, however, with that of belief and unbelief in the sense customary in Holland, that one is justified in considering many of those who have faith in the church as unbelievers, and many who have no faith in the church as exponents of belief. The religious struggle of by-gone centuries, which took a different view of Dutch national history – for a long time resistance of the Netherlands against Spain was regarded as rebellious by the Catholics – has lost its momentousness in our day, because now the most powerful forces no longer clash over religious faith, but over the form wherein the individual shall have a right to the goods of the community and where the moral basis is what counts, no matter whether this form shall be individualistic, allowing preferential judgement, or whether it shall show solidarity, proclaiming the rights of all to a share in the material and spiritual goods of of this world. In this struggle those who call themselves believers in the traditional sense, are overwhelmingly on the side of conservatism, whilst among those who call themselves unbelievers there are many who would sacrifice everything to make the great cause of social justice be succesful, and who thus display the essential hall-mark of what belief and faith have always been, namely that a man shall allow himself to be guided, without reservation, calculation or fear, by some principle. When they, who through tradition lay claim to piety, but do not display unreserved self-surrender and vice versa, should one then not say that belief has turned into unbelief, and the reverse? The well-known utterance of the clergyman Niemöller that the unbelieving West is trying to keep the believers of the East at bay and to destroy them through arms of an unprecedented power, is a clear illustration of the shifting of content and function that is taking place in the world between belief in the traditional sense and belief in the actual sense.

It is not really possible to leave these things out of consideration, when one is examining the responsibility of the research-worker in the cultural sciences as a matter of principle. The teaching of history and of the knowledge of present-day nations will vary according to whether one considers the lead, Western civilisation has attained in the last four centuries, as definite and as proof of spiritual superiority, or whether one recognizes that the relationship of Europe and its Western descendants to the rest

of mankind should be upheld by the acknowledgement of everyone having equal rights and feeling respected for each other and
whether one faces the question squarely that the practical consequences will be of this realisation. As long as the idea of colonial
world dominion was covered by the spiritual claims of Christianity to being the absolute truth, it was possible for the white race
to look upon itself as divinely chosen, both materially and spiritu-
ally and it could feel itself called upon to bring the light of truth
in the darkness of Asia and Africa, while at the same time robbing
them of their freedom and exploiting them. This self-assurance,
in which lust of power and idealism were unequally mingled, is
making way in our time for the painful experience that the people,
who have been subjected, are laying claim to freedom and that
seen in the light of how we conquered and exploited them,
they do not feel the respect for our spiritual benefactions that
was expected by those who brought them sometimes in good
faith and sometimes in bad. How would it be possible for this
fundamental shifting of the rôle the West takes in the world, to
remain without influence on the science of geography. How would
it be possible for it not to support the more objective idea and to
consider it confirmed by the course things have taken, which
fifty years ago scholars like Snouck Hurgronje, van Vollenhoven
and others of their generation put forward and which, notwithstanding their authority, aroused so much annoyance and resistance. We have the privilege of living in an age in which a new
universalism and a new solidarity are beginning to manifest
themselves, an age in which the rottenness of the old slogans is
being found out, those slogans which still sound good but which
no longer have any life in them, because betrayal of human digni-
ty and mental self-conceit lie at their roots. It may be so that the
new values that are placed over against the old ones are really
those old values, only set free from the grip which prevented
them from having effect. This understanding cannot make the
conflict between the old and the new Western world any the less.
That which is really a new faith, will be greeted by the diehards
as unbelief and deterioration and conversely the old faith will
have to go on hearing the reproach that it has in fact become a
form of unbelief.

It is clear, that the researchworker in the "cultural sciences" finds himself right in the middle of all this tension. This is only seemingly not so as long as he allows his attention to be caught by small questions, far-removed from the circumstances in which he himself lives, and which since they belong to the immense field of the scientific reconstruction of the past of mankind, require the intense concentration of the seeker after truth. There are many such questions and the work spent on them should be appreciated.

To stick to examples out of my own field of teaching, it is undoubtedly not of decisive importance whether the year of Plato's birth is put in 428 or 427, or whether the laudatory speach that Plato lets Alcibiades give on Socrates represents the actual relationship between the two or whether it idealizes it. Principles of moral judgement are, however, also required for such small problems. An intuition that is fertile will always be a gift from the gods, methodical justification, arguments that will hold water may not be lacking afterwards, so long as it serves truth and is not done only to please oneself or other. In the cultural sciences there are an endless amount of questions of a purely factual order which must and can be solved sine ira et studio. This is a requirement not only of objectivity, but also of concretion: an idea of human reality cannot be built up out of generalities. And as regards objectivity, the research-worker in the humanistic sciences can only rejoice when he contributes in increasing its domain. It is true that his joy is not free of its dialectic antithesis: as long as a problem is not solved, it moves men's minds in another and more turbulent way than when the solution has been found. There is an element of rigidity, of hard-and-fastness that makes people prefer the tension of seeking above the steadiness of knowing. The pursuit of knowledge of the last two centuries leaves little room, though for the oppressive feeling that sometimes emanates from an incontrovertible truth. On the contrary Lessing's well-known words have become characteristic for the whole of research. There is no longer any question of observation and enjoyment of the past on a selfcontained image, being the aim. For are not all the "pictures" of the past a sort of construction that is never finished which all arises out of action and remains connected with action? This pragmatic basis remains, even after the original, natural involvement of the cultural sciences with life has been understood and corrected as being limited and limiting. It is not the pragmatic part that disappears but the limited part which consisted in the research worker at first making his own "lebensraum" guilelessly absolute. Breaking down egocentric and particularist thinking means paving the way for altruistic universalism of a true solidarity, in which each limited domain of the past has been overcome and co-ordinated.

The investigator of the humanistic sciences is in the first place, although not solely, a critical judge who allows objective truth to remain in force, when confronted with an incorrect image of it. What irrationalism teaches a b o u t various gradations, therein sometimes even gaining the approval of some rationalists, namely that criticism only tarnishes the appearance of human things, is not true. For criticism may go with an enthusiasm that is not accompanied by any blinding or illusion. I t m u s t n o t b e t h o u g h t t h a t t h e r e c a n b e n o h u m a n g r e a t - n e s s b e f o r e t h e b a r o f l e a r n i n g . The image that was crushed by criticism, was mixed. Reason which pulled down itself builds, in the place of the old image, a fresh one of a stronger make that can pass through the ordeal of criticism in form and substance. The coming and going of public favour does not touch the images that rest on the pedestal of the lasting memory of mankind, elevated above all temporary revilement and oblivion. This is the temple of the great benefactors of whom Auguste Comte spoke so sanely, so critically and at the same time so animatedly.

For this reason the responsibility of the investigator of the humanistic sciences is a double one. His criticism and causal analysis that spares nothing end in worship and inspiration. He is judge over the indisciplined passions as well as keeper of the temple in which they reside who stood out so far from the rest of their fellow-men that they were like gods. History has brought forth gods in the old Hellenic sense in such mythological and historical figures as Prometheus, Socrates, Jesus, Marcus Aurelius, Eckehart, Spinoza, Kant, Hegel, Marx and so many others. As a keeper of their temple the student of the history of culture has a mediating task. Without himself being one of them, nor yet belonging to the ignorant mass of men, he calls for respectful silence and on the one hand wards off the desecration threatening in fantastic imaginings and on the other hand he rouses the masses out of their blighting sensuality. As educator in pure humaneness, i.e. a harmonious mingling of sober-mindedness and inspiration, he cannot remain indifferent to the evil man consciously or unconsciously does to man by the plea of truth. In his fight against evil he meets the natural science research-worker whenever the latter realizes that it is contrary to his sense of responsibility to render services which are paid in tempting monetary rewards or in credit, but whose tendency is to further the economic benefits of certain groups at the cost of others or to prostrate the adversary through inventions of destructive force. As long as the instruction of the masses via teaching, the newspapers, the wireless and the pulpit is employed for arousing feelings of distrust, aversion and fear, the sense of responsibility of the scholar will prompt him to take a stand against this in the name of brotherhood and humaneness.

Thus in our time the responsibility of the investigator of culture has come to the fore as the morally essential part of his work. In former centuries the science of man, society and history may have been much like a game to those who felt the need of filling the void, social privilege sometimes engenders. In our time there is no longer any question of there being a right to such a pastime, and it would even be erroneuos to call it harmless because it is peaceful and non-aggressive. This game is not harmless in that it seemingly stays outside the conflicting forces in the world and yet in the last resort chooses the side that offers the best changes for its continuation. Neutrality is not objectivity, to keep out of a conflict may be a formal right, it is not a deed of justice. The final say in the pursuit of the humanistic studies is consequently the sense of responsibility, in other words the attitude one adopts. Whilst in others the position they take up is collectively determined and is inclined to call its natural necessity freedom, it becomes r e a l freedom for the investigator to the degree in which he allows his feeling to be formed by his understanding and allows it to be released from the unaffectedly blind servitude to his own welfare or limited aims.

The consciousness of his responsibility enables the researcher to attain freedom. This freedom is not outside or against reason, but it is within and exist by virtue of it. Whoever recognizes this

11

will no longer feel any distrust towards knowledge and criticism. The profound truth of Spinoza's words that the perceptive a-wakening of our nature and that of fellow-men is the path to inner freedom, and those words of Hegel's to the same effect: D e r B e g r i ff i s t d a s F r e i e, correspond to his experience and show him the way. A sense of responsibility in the researcher only serves a useful purpose if knowledge and action are virtually not separable. They can only be so for whoever continues to think formally and for him who does indeed realize their fundamental unity but who cannot demonstrate this unity on account of his particular nature. This is a case where Hegel's term "the unhappy consciousness" is applicable. But there are also people of happy consciousness – we do not mean the self-centered enjoyment of narrow minds –. These admit that insight implies action.

DEBATE ABOUT PROF. H. J. POS' PRE-ADVICE ON "RESPONSIBILITY IN THE HUMANISTIC SCIENCES"

by G. Gonggrijp

It is not my task to point out all that I gladly subscribe to in Pos' pre-advice or what had my full concurrence, but I must try to say in ten minutes what objections I have against his utterances. In so doing I must confine myself to what is most important.

The investigator of the humanistic sciences is responsible for the dissemination of knowledge in a spirit which does not only show no respect for making what is limited absolute, like chauvinism, imperialism, racism and all fanaticisms, but which expressly denounces and fights them. Excellent! "Neutrality is not the same thing as objectivity". Exactly! "The final say in the pursuit of the humanistic studies is consequently the sense of responsibility, in other words the attitude one adopts". Exactly!

Now Pos believes that the student of the humanistic sciences should be aware that a choice should be made between an old and a new faith, between the belief in the value of the moral tenability of our western world and the value of the East, by which Pos, as appears from his quotation of a saying of Niemöller's means the world whose power is centred in the Kremlin. There in a current sense the new faith is supposed to inspire men's minds. He characterizes the contrast between our world and that of the Kremlin in the following way. Here in the West we have the form wherein the individual will have a share in the goods of the community in an individualistic way, while permitting privilege and the continuance of the rift in the community; while in the East, in the world of the Kremlin the form of society rests on solidarity, the right of all in a share in the material and spiritual goods of life.

The characterization of our western world as being individualistic is already greatly out of date in numerous instances and untenably simplified. For it is here in the West that there is abundant opportunity for individuals jointly to strive after goals in various fields of our economic and social life, for instance in the claim for wages or for realizing spiritual and political ideals, all

of which can be done in absolute freedom. One should not gather
from the foregoing however, that by this I mean to say that I
object to the individualistic features of society as we know it.
On the contrary! Wherever necessary, we must carefully safe-
guard and protect it from being suffocated and stifled by too much
solidarity, by the communal or social principle or whatever one
wishes to call it. For it is true that man is a communal being,
but I believe, and on good grounds, in the creative powers in the
economic field and in those of science and art of the free individual. Woe to the society in which the individual no longer enjoys
the freedom of making his own way clear where necessary and
of defending his own thoughts and inventions against the o-
pinions of the group or groups to whom he belongs, or against
those of the state. That there is often a great deal of co-operation,
and that this must be so, in science, politics, economic life and
art is self-evident.

The principle prevailing in the new world, in that of the East,
Pos calls solidarity, the principle he himself believes the one to
choose. I can understand that he chooses this word and does not
speak of communism. Not only has he objections against communism, against the practices of the Kremlin system, but it
would sound much too crude if he were to say: I choose the
Kremlin brand of communism.

Nevertheless I regret that he does not use the word communism
or Kremlin communism. For in science we must be as clear as we
can and choose terms that leave no margin for doubt but call the
things what they are. And Pos undoubtedly chooses Kremlin
communism.

He cannot choose solidarism as he calls it and not Kremlin
communism. Pos chooses, he chooses between the two worlds,
between which everyone has to choose in this period of time.
Because neutrality at the present day is immoral and untruthful.
As untruthful as it was in time of the occupation to say I am not
for Germany, nor for Holland, I am neutral. Pos is not neutral,
and that I appreciate! But one cannot say, I choose the East, the
Kremlin system, but not really, because I choose solidarity,
altruism, the solidarity of universalism. Whoever chooses the
West, chooses the West, our world, including all the things that
are not good in it as well. And that person cannot say I choose

the West, but not really because I choose the sermon on the Mount. That would be untruthful. Now the important thing is that whoever chooses the West can fight the things he thinks wrong in perfect freedom. But he who chooses the Kremlin world, and who helps the Kremlin to power, cannot fight the things he thinks wrong in perfect freedom. That is a real difference!

One cannot say I choose "the East", but without the forced labour camps, without the shameless exploitation, without the lack of freedom, without the false propaganda and without Soviet imperialism (Soviet imperialism is the continuance of Russian imperialism for which Karl Marx warned the world so forcibly in the most topical of his writings).

Pos wants solidarity. My objection to that term lies in the fact that the reality of the totalitarian system that Pos chooses is a bitter and cruel parody of solidarity.

Solidarity has been realized to a much greater extent in our own Dutch society for instance than in the Soviet Union.

The grave thing about all this is that Pos, who naturally wishes freedom for science and for those who practice it, chooses a system that is totalitarian, and this totalitarian system is not only dangerous precisely and mostly for the humanistic sciences, it is on that account really irreconcilable with it.

I consider this a grim state of affairs. And I say this because I am very well aware of the responsibility of the student of science.

You paint things in black and white, Pos will say, that is scientifically speaking not allowed. It is not seemly at such a congress as this. No! I must not discuss the bad, the less bad, the less desirable, the fairly good and the good in the Soviet Union in varying shades of grey. That is not possible at this congress nor can it be done in ten minutes. I must think of principles. And that is why I say clearly and plainly and in no uncertain tones, as against what Pos says, that I call choosing the East and at the same time wishing for freedom in science, inconsistent. We experienced this sort of thing during the occupation. National Socialism and the Kremlin system correspond in this also that they both of them, by their very nature of totalitarian systems, and resting on the powerful machinery of a secret police service, that in practice is above the law, murder freedom, including the freedom of the pursuit of science.

Pos quotes an utterance of Niemöller's with approval in which the latter says, that the unbelieving West tries to keep at bay the believers of the East by means of unimaginable power. Completely incorrect! We do not need our mighty weapons for keeping the believers of the East at bay. We could manage their belief alright! If the iron curtain, whose most important function it is to prevent the truth about Russia from penetrating into the West, and especially from preventing the truth about the West from penetrating into the Soviet Union – because that would be particularly dangerous for the rulers in the Kremlin! – if the iron curtain could be withdrawn then the power of the Kremlin would soon start tottering. Our powerful weapons are not necessary against the believers of the East, but against the weapons and the armies at the disposal of the Kremlin . . . .

The Chairman: The time is up.

SUMMARY OF PROF. H. J. POS' ANSWER TO PROF. G. GONGGRIJP

In his discussions with Wertheim and now with me, Gonggrijp has transgressed against the rules fixed so wisely by the Committee of this congress, that politics should be eschewed. This morning he said, and it was tolerated by the chairman, that he applauds the arming of the United States since they stand for our safety, and he lays at my door that I am supporter of the Kremlin and he has twisted the meaning of my words. The non-political, but psychological question that I put was why it is that in the East faith is so great and that in the West it is fear and the feeling of discouragement that are so great. And further as a thinker on culture I set the moral value of an economic system of free competition against a system of solidarity as regards the meeting of needs. Gonggrijp is an enthusiastic supporter of a system which to my mind is as unreasonable as it is conflicting with the Christian religion. In order to acquire an opinion on this it is not necessary to know much about Russia. The daily study of the great European thinkers from Heraclitus to Marx and Nietzsche leads to such a view. The glorification of the armaments of the mightiest country in the world does not take count of reality. It's real motives are complicated and not so philanthropic as Gonggrijp's egocentric thinking would have us believe.

He can know better than I that in an economic depression armaments orders are the only expedient for preventing a catastrophe. However, this solution creates the danger of fresh catastrophe.

DEBATE ABOUT PROF. H. J. POS' PRE-ADVICE ON "RESPONSIBILITY IN THE HUMANISTIC SCIENCES"

by K. KUYPERS

In his argument on the responsibility of the humanistic sciences the pre-adviser sharply contrasts the view of the task of a rational science of history with that of Romanticism and the irrationalism and subjectivism that go with it. Notably in Romanticism he sees a return to a pre-critical conception of science. He directs himself particularly against the historical view brought forward by Romanticism which involves truth itself in a process, and would thereby, as he believes, ultimately obliterate the distinction between truth and belief.

Over against every form of Romanticism he puts the critical view of history that is guided by the ideal of a new universalism and solidarity.

One can hardly say that this view of history is generally accepted by contemporary historians. This view naturally does not stand or fall by this. We consequently do not wish to make a call on this in any way, but we will rather ask ourselves on what assumptions this view that the pre-adviser advocates are based and in how far they seem tenable to us.

In the first place we note that the pre-adviser consciously falls back on the critical attitude towards the past as this was evinced in the eighteenth century, the age of the Enlightenment therefore. Just as did the age of the Enlightenment so also does the pre-adviser base his critical attitude solely and only on reason. On these grounds he too sees in the investigator of the humanistic sciences before everything else a critical judge who can only usefully fulfill his task as historian if he performs it while basing it on the ideal of universalism and solidarity, i.e. the idea of social justice. By the side of this soberly critical task which puts aside all delusion, dazzlement and illusion and in its stead creates an objective picture of reality in the service of truth, there is still the possibility that only human greatness is able to vindicate itself before the bar of science and is able to arouse true enthusiasm.

When we review the whole argument of the pre-adviser closely, an argument which to retain the mode of expression of the esteemed pre-adviser, is somewhat of being a sharp summing up with respect to the whole part of Western civilisation on the grounds of an ideal for the future and of faith, then we believe we can discover a discrepancy. For does not the pre-adviser himself remark that under the pressure of circumstances the view about the sovereignty of the state has changed. This doctrine, as long as it corresponded with legal reality, can no longer be recognized as such, because the independence of the state no longer exists in that sense on account of the historical development of the relationship of the states amongst each other and it cannot be sought after any longer either. He admits that the new values of universalism are not entirely new, but rather the old ones now freed, however, from the grip which hampered their working. And finally he mentions with concurrence – and this is decidedly important – the giving up of the view about a permanently attained truth. And this not only as the pursuit of science in general but especially about our knowledge of the past also. The past is never a closed case of definite form. From this it appears that the pre-adviser himself acknowledges that truth is part of a process of awakening, and that historical action should always be grasped in connection with a historical situation. As it seems to us this also means that this action should first and foremost be judged in relation to a special historical situation.

We shall leave out of consideration that the distinction made by the pre-adviser between the natural sciences and the humanistic sciences, as also the central place he assigns to history, has its origin in Romanticism and the historical school. What matters here however is the implicit recognition that the rational mode of viewing history sees in history a process of the awakening of truth, and that historical action should in the first place be judged in relation to historical configurations. The historical thinking of the nineteenth and twentieth centuries has taught us how to handle this concept of relativism, and at the same time how to learn to understand the past in a manner and with means unknown to the eighteenth century. The clear understanding of the past which historical thinking is able to give is richer and profounder than the critical judging and rejection of the eighteenth century. We believe we must look for the cause of this inner inconsistency in the fact that the pre-adviser links the eighteenth century rationalism of the Enlightenment with the concept of relativism, which has its origin in the historical thinking of the nineteenth and twentieth centuries, without to our idea sufficiently realizing this connection that he makes. He criticizes the historical thinking and investigation of the nineteenth century and our own time from the standpoint of the eighteenth century belief in reason, and on the other hand he criticizes the past of European and of our own national history with the means that this historical thinking has learnt to handle, namely relativism. Because the pre-adviser does not explicitly admit this, a striking one-sidedness arises which consists in his only being able to see in the historian nothing, if not the judge and critic, for which reason he makes the task of student of the humanistic sciences turn on historical criticism. It is true that as a basis for historical criticism he now puts the ideal of universalism and solidarity.

However important this basis is, it can never be set as the only aspect of all historical research, nor can all other aspects be made subservient to it, and finally nor can all the significance of historical research be measured by that alone. For the result is that the pre-adviser is inclined to consider a large part of what historical research workers actually do – we are thinking for instance of historians of language and of art, but also of philologists – as valuable, that is to say in the most favourable instance, but not as of real current interest, or he even degrades it to a more or less pointless pastime, in which the research worker is kept occupied with curios, his social position enabling him to do so.

This one-sidedness, partly a result of the inner inconsistency that has already been pointed out, brings along with it that the pre-adviser, in his meeting with culture as a living historical reality, and with cultural science which makes reality its object of investigation, only sees the relation of the dependant and his critical judge, who himself is seated on the throne of unshakable reason. On the other hand it cannot be emphatically enough pointed out that there is a close relationship between the historical understanding of the past and the cultural life in which the historian himself participates. If a united Europe should ever become a fact then the historian will also look upon the past of the European states with different eyes. The understanding of reasonableness is itself not absolute but it develops in cultural life itself. This is usually not effected by the historian, but more possibly by other sciences, persons or forces in the cultural life. It is not the historian who has substituted a scientific picture of the world for the former mythical one, but the philosopher and the critically thinking individual. The task of the historian is therefore much more modest than that of assayer of cultural life, and it is also much more positive. The historian is not responsible for the general truth of our idea of the world, nor yet for social justice but he is responsible for the historical truth i.e. for the truth about the past, for the truth for instance about the mythical idea of the world. This historical truth, however is not a fixed quantity. The student of law and the economist are much rather responsible for the social justice that is actually brought about in society, than the historian. According to the romantic idea of the humanistic sciences these too should only be looked upon as historical research workers, since according to this idea history is the pivot.

Now as regards universalism and solidarism, which the pre-adviser wishes to be the basis for historical criticism, we shall have to confine ourselves to a few remarks which, however, join on to the foregoing. The human mind is not dependent on the historical situation in which it finds itself. To such a degree that it can anticipate an idea like that of universalism, but for its realization he is dependent on factors which are not without more under his control.

A moral conviction, however essential it may be, a basic condition in fact, is yet not sufficient. For the realization a special configuration of cultural life is required. The idea of mankind as being one whole is a very great idea, it was anticipated by the stoics and by Christianity, but it remained an utopia, if only on account of lack of contact and even of knowledge of each other's existence. Only with the development of physical science and technical science did the contact come about, and the interconnection and the real possibility of equality, and was universalism as a moral claim brought to the stage of real current interest. Because it cannot be said that the means were consciously at the service of the furthering of humaneness and brotherhood, or that it is so to-day, but that the contact and the interconnection of

mankind did not come about until then is a fact. In this connection one should measure the position of the bearers of Western
civilization with historical standards, and not afterwards with
the supposed absolute standard of reason, which is no better than
the level of the present day historical development.

The pre-adviser rightly stresses the point that speaking of
freedom and restriction in science can only be meant in relation
to the students of science and learning. Wrapped up as he is in the
consideration of the moral responsibility of the students of the
humanistic sciences. that of freedom is more or less relegated to
the background. Only at the conclusion of the pre-advise is
freedom spoken of, and then in the sense of a principle arising out
of reason itself, making itself felt by virtue of it. We should especially like to see the responsibility of the students of the sciences of man brought within the scope of the maintenance and
protection of spiritual freedom as a possession that has been attained with difficulty and that has not only been threatened in
the past by powers that are totalitarian or not as the case may be,
but that is still so threatened.

We shall be glad to hear what the pre-adviser thinks of our objection, namely that in his argument we feel the lack of a plausible
integration of the uncompromising concept of reason prevalent
in the eighteenth century with the development of the historical
thinking of the nineteenth and twentieth centuries, that is in
agreement with the present day dynamic idea of the world. In
our opinion historical insight pre-supposes transcendental criticism or at least includes it, but it is by no means absorbed by it.

SUMMARY OF PROF. H. J. POS' ANSWER TO PROF. K. KUYPERS

The criticism uttered by Kuypers breathes the well-balanced scholarly spirit that might be expected of this thinker. To my idea it entirely answers to the intentions of the Committee of this congress. A serious objection to the against his opinion, as if the distinction between the natural and the humanistic sciences and the constituting of the latter were of Germanic origin lies in the fact that the Enlightenment did not only bring forth historical criticism but also the universal interest in mankind including extra-European cultures (China, India). The eighteenth century has also brought forth master pieces of cultural histories, like Voltaire's "Le siècle de Louis XIV" and "Essai sur les Moeurs", Montesquieu's "L'esprit des lois" and Condorcet's "Tableau des progrès de l'esprit humain". The German contribution to the humanistic sciences is over-estimated by it being brought into connection with Romanticism. That the unity of mankind in the shape in which the Stoics and Christianity advocated it remained an utopia is correct. However it is today of greater current interest than formerly, on account of the fact that the idea of "one world or no world" has become a vital necessity. As against this necessity a united Europe is in so far of subsidiary importance in that it can be of no significance if a greater antagonism were to remain in existence round about it.

THE RESPONSIBILITY OF THE SCIENTIST

by A. G. M. VAN MELSEN

1. *Introduction.*

In the foregoing studies it has been shown from various points of view that freedom forms part of the very essence of science and knowledge. Freedom, however, does not mean arbitrariness. Freedom means that the way science is bound can never come from outside but that each science must itself determine to what it shall be bound. It can and may only bind itself to that which it has determined shall bind it, namely its object and its method. That is to say that whatever any branch of science considers true is on the one hand determined by the given facts and on the other by the way in which man can best encompass these facts in a homogeneous system of concepts.

Now what is "given", the actually existing reality, falls apart in various fields. Besides this each field can be viewed methodically in more than one way. It has appeared to be of advantage to keep these fields and methods separate, so that gradually a whole spectrum of sciences has arisen, each with their own object and method.

In what is to follow interest will centre in the so-called exact sciences, although the subject here under discussion naturally does not entail that the method we shall follow will be that of the exact sciences for the simple reason that terms like "scientific freedom" and "responsibility" are beyond the range of the exact sciences which of course does not mean that they are beyond the scope of the *student* of the exact sciences. In the pursuit of science he bears the responsibility because every specifically human activity because it is a free activity brings responsibility with it, and the more so in proportion as the activity in question is directed towards the higher and more central human values. It is freedom therefore that brings responsibility with it. Where there is no freedom there can be no responsibility. For this reason responsibility is only to be found at the human level. Only when an action is free, i.e. the result of the free self-determination of a creature that knows what it is about, does responsibility come into the picture. For in order to be justified a free decision must be in agreement with the standards that apply to for the action in question. Now standards must be *recognized* if they are to standardize an action. This has an important consequence respecting the responsibility resting on the student of science. For while studying reality science at the same time discovers certain standards for human actions. Physical science for instance teaches us how matter must be worked. Through this standardizing function, which is an inevitable concommittant of science, the student of science also bears a certain amount of responsibility for the action of his fellow-men, who look up to his scientific authority. Of course it remains true that the more fundamental ethical norms of human actions cannot be determined by positive science, for the simple reason that they lie outside its domain. But it is equally true that science has an important function in the correct application of these norms. Erroneous scientific theories can thus have results of a far wider bearing than an incorrect theoretical concept. This naturally also has its repercussions on the responsibility resting on the student of science, a responsibility which is all the greater in proportion as the science in question is more directed towards higher human values.

When considering the responsibility which comes to rest on the student of science it seems to us useful to distinguish three fields which each brings along its own problems. These three fields comprise: 1° the pursuit of pure science, 2° the spreading of science, 3° the application of science. Naturally the boundaries between these fields cannot always be sharply drawn, they are, however, sufficiently clear to justify the division we have just given.

II. *The pursuit of pure science.*

Responsibility in the field of pure science raises the least difficulties because in this field it is almost solely determined by the requirements of science itself. The pursuit of pure science is after all no more than the fulfilment of a fundamental urge in man, namely the striving to understand reality consequently, reality itself and the methods to study reality and nothing else determine the standards of scientific research.

Every deviation from these standards impairs science itself. So the only thing for which the student of science is responsible consists in being able to prove his statemento, i.e. to demonstrate that they actually are in accordance with reality. To this we may add, that in expressing scientific opinions, the certainty with which they are expressed must naturally also be justified. It is scientifically unjustifiable to represent an opinion as being certain when it is not certain but only probable.

The foregoing is self-evident really to every serious student of science; pure science is only justified through its scientific standards. In accounting for his scientific results a student of science need not and may not therefore trouble about their possible consequences, as for instance whether they are of economic or moral use. Considerations of utility of all sorts may perhaps prompt him to start a given investigation, but they can never give the justification of the scientific opinion itself. One might therefore distinguish between an external and an internal responsibility, There may be special reasons that spur someone on to pursue this or that branch of science or to start with this or that investigation, which each bring along their own responsibility, but this responsibility remains external to science itself. Within science the only responsibility is veracity. If the results of the scientific research do not tally with the considerations that were the reason for it, there can be no doubt as to what prevails.

Perhaps it is needless to point out that every scientific view is a good, for it means the enrichment of man. For this reason there is not a single datum of which it might be said that it would have been better if man had never known of it. In this connection I am thinking especially of the general sense of oppression that took hold of mankind when nuclear energy had been discovered. Oppenheimer tells how after the first (experimental) explosion he prayed a line from ancient India: "I have become death, the destroyer of worlds", whereupon Bridgeman answered: "if any should feel guilty about this it should be God. He put the facts before us as they are" [1]). This answer was correct. Considered in itself the knowledge of nuclear reactions is a good, because it deepens our insight into the nature of matter. The scientist may shrink at the thought of the possibilities there are for abuse, but he is not responsible either for the forces of nature, nor for their abuse. As a student of science he is only responsible for a

---

[1]) "De Linie", April 24, 1954, p. 1.

correct understanding of those forces of nature. No student of science as a matter of fact can foresee to what use a certain result of science will lead, consequently his specific responsibility as a student of science cannot lie in that direction. The use that is made of given scientific results does, of course, call up other responsibilities, but we shall discuss that later. We must pause here for a moment however, to consider something else.

The fact that every scientific insight is a good, does not mean that we must strive after it cost what it may. Thus it may very well be that certain experiments should not be indulged in on ethical grounds and that for that reason they are not justifiable. These restrictions, which do not affect all that is worth to be know but which do affect the means for gaining knowledge, occur in many branches of science, and naturally the most in those that have a bearing on man. The ethical restrictions just mentioned are the result of the fact that scientific activity, being an aspect of human activity, remains subject to moral standards which hold good for every human activity.

III. *The diffusion of science.*

As regards the responsibility which the scholar bears in connection with the diffusion of his knowledge, it seems desirable to distinguish between the diffusion that is exclusively confirmed to fellow-experts and that which also extends itself to non-experts, by which I have more especially in mind the diffusion of science among the general public.

In regard to the diffusion of science among fellow experts it should be remarked that this diffusion belongs to the very essence of scientific research. For science is a general human affair, truth is never a thing that holds good for only one person, science is inter-subjective, because it has to be objective. In view of the fallibility human knowledge one might even state that the responsibility which a scholar bears in regard to his branch of knowledge brings along with it that he should subject his scientific views to the judgement of others so that he may thus attain to greater certainty. In addition to this it should be remarked that a certain amount of one-sidedness is attached to every scientific method. Exchange of data and views is therefore an indispensable condition for true science. So in general it can be said that the responsibility which the scientist has with respect to diffusion of science among fellow-experts or among students of other branches of science that are more or less related, is of the same nature as he has with respect to the pursuit of science of which we have spoken above, for which reason we need no longer pause over this point.

With regard to the diffusion of science outside the circle of fellowexperts, matters lie somewhat differently. Some particular factors come up for discussion in this connection which justify our pausing here a little longer. Let us start by pointing out that the scientist, in his capacity of the exponent of man's searching intellect, is partly responsible for the views of his fellowmen. His views must benefit his fellow creatures in proportion to their powers of comprehension, and in this especially lies a grave responsibility. What may lead to a better understanding of a thing in a well-trained mind may easily result in misconceptions in a mind that is not trained. In so far as such a misconception lies in the field of purely professional science it does not matter so much. That a schoolboy on his first acquaintance with the atom theory should think of atoms as small unchangeable spheres is probably a more or less inevitable stage in the development of his views on science. Such a misconception can later be easily corrected. The decisive point regarding the responsibility lies here in the question as to whether the initial misconception was inevitable and whether it can be corrected at a later stage. It is therefore a didactic problem.

A much more serious problem regarding responsibility arises, however, when the information that the scholar believes he should give his fellow-men exceeds the boundaries of his own special field. Naturally we are not thinking of those extreme cases in which a person goes entirely beyond the limits of his special subject; a physicist for instance who should start talking on economics and who in doing so tried to cover his opinions with his authority as a physicist. Although such cross cases do occur we need not discuss them, they are scientifically injustifiable, and the general wisdom of the proverb that tells the cobbler to stick to his last offers a sufficient counterbalance against such escapades. There are more subtle infringements of the boundaries, however, which are often inevitable and which on account of their subtility are not noticed so quickly and are therefore more dangerous.

Let us start by at once stating that the boundaries between the various branches of science are by their very nature not very definite. This holds good for the different positive sciences a-mongst each other and also for the positive sciences on the one hand as against philosophy and theology on the other. Then it must also be remarked that the student of science rightly feels responsible for misconception outside his special subject, in so far as those misconceptions to his mind are the result of a wrong interpretation of his own subject. In this way a physicist may have the conviction that certain philosophical theories result from physical theories that are out of date. A philosopher may be equally convinced that certain ideas current among physicists originate in incorrect philosophical interpretations. The same may occur with regard to theology.

Such a situation or fancied situation induces the student of a given branch of science to go beyond the boundaries of his own subject, which naturally constitutes the danger of injustifiable extrapolations and distortions. And it is then especially that the danger of misunderstanding arises, engendered in immature and unscientific minds, to which we alluded in the above. For when the student of a given branch of science, in the opinion that his subject can give an important contribution of a philosophical nature, proceeds to expound his theories to a general public, then he can hardly avoid to arouse misunderstanding, obliged as he is to popularize on all sorts of points, than. Since he cannot use his specific technical terms, he is obliged to fall back on general terms and concepts, which are not really suitable for rendering his meaning. The result is often a gross misunderstanding. The history of science, alas, presents many such examples. One need but recall how in the previous century physicists for instance spoke about free will, which they rejected on the strength of the general determinism shown by the natural processes observed in physical science. Other examples are the theory of evolution, the theory of relativity, psycho-analysis and the quantum theory. To give a contrary example we may point to the injustifiable way in which many theologians of the seventeenth century went on ad-hering to an idea of the universe that had long become out of date. The whole wretched controversy between faith and science, which made so many minds miserable in previous centuries, must largely be put down to the injustifiable infringements of scientific boundaries, which we pointed out above. Or rather, it is not so much the fact of these infringements themselves that should be held liable. For we have seen above that they are more or less necessary. What should be especially held liable is the fact that the scholars in question were too little aware of these infringements. Within the framework of this study about the responsibility of science the great responsibility must therefore be stressed which the student of science bears when he believes he is obliged to go beyond the boundaries of his own subject for the sake of enlightening the general public. And his chief responsibility is probably to be aware of the limits of his own field of knowledge and the theories that are there applicable. He must particularly realize in how far that which he puts forward really springs from his subject and not from a particular philosophical conviction in the light of which he interprets the data of his branch of science. No positive science is able to vindicate its philosophical suppositions, for which reason every science in principle retains something unsatisfactory intellectually, which compels especially the best and most eminent students of science to resort to philosophical speculations without their always being aware of the different character of those speculations.

IV. *The application of science.*

When we look at the history of science, we are struck by the fact that science and technics approach each other ever more nearly. The first rational thinking, as we see it revealed in the Greeks, is a reflection on the fact that there must be a certain rationality in nature. Multiplicity must be based on unity, mutability on immutability and qualitative variety must be capable of quantitative comprehension. Considered in themselves these philosophical speculations naturally do not give us physical science in the modern sense of the word; they do, however, pave the way for it, because they show that there must be general and immutable laws hidden in the multiplicity and immutability of the natural phenomena. The knowledge of these phenomena for the time being remained wholly based on practical experience. Between theory and technics, the art of manipulating things, there remained a great gulf. This gulf became smaller in proportion as specific physical scientific theories, showing an understanding of special phenomena, came to the fore by the side of the general philosophical theory about the rationality of nature. With this growing understanding the gulf between theoretical and technical knowledge diminished. This understanding gave direction to technics and this in turn opened up the possibility of penetrating deeper into matter. Aristotle has characterized the process of knowledge as a process in which the knowing subject identifies itself with the known object. To know means "to become" the known object, because the intellect takes on the essential form of the thing known. In modern technical science we see as it were an illustration of the reverse of this. Thanks to the understanding man has gained about matter, thanks to intellectual penetration of matter he is able in his turn to realize his ideas in matter and thus to make matter subservant to his human ends. Not that matter has lost its strangeness and otherness in regard to man, for man can only control matter by understanding it, by putting himself in the position of matter as it were. Only by taking matter and its laws as his starting-point, by intellectually "becoming" matter, can man control matter and thus "humanize" it. The "humanization" of matter is expressed in modern technical objects of utility, in means of transport and communication, in machines, and – to take science only – in making use of scientific instruments, without which modern science is no longer possible.

The observations we have just set forth, are meant more especially to show how closely science and technics have become interwoven, the one is unthinkable without the other and inversely. In this way science has taken a new responsibility upon itself, the responsibility of maintaining this world of technical science which has become a part of our human world, and to perfect it further for the benefit of man. This responsibility appears from the changed position of the universities. Once they were above all institutions for intellectual thought and reflection, without any direct influence on the daily material course of the life of society, whereas to-day they are an indispensable constituent of that life. If scientific forces were not to pour into present-day society abundantly, it would go under because the whole of its machinery cannot be kept going without scientific activity.

The prayer: "Give us this day our daily bread" has to-day above all become a prayer for scientific processes.

It is as clear as noonday, however, that the responsibility of the university does not end with the provision of scientifically trained "technical experts" who know how a certain goal may be attained technically, or who are able to open up fresh technical possibilities by the aid of ever advancing scientific research. For the whole of technical science is only a means to an end. The human purpose for which these means are employed remains of decisive importance. Medical science is today able to prolong life and to prevent or cure numerous diseases; and physical science provides man with the means to encircle the whole of the earth with his physical presence, but what does man do with this longer life, and how does he fill his increased possibilities in living? These questions are alarming, and that the student of science is especially aware of this proves that he realizes that the responsibility of science taken as a whole goes further than the care for insight into the positive sciences alone. A vestige of the old cultural care of the university as it used to be still lives on even in our day, and probably a little more strongly to-day than in the times that lie immediately behind us. The responsibility resting on the university should indeed reach further than simply the practising and teaching of positive science, it is also its responsibility to make its students grasp the fundamental human values to the best of its ability. This last is said advisedly because it is certainly not exclusively the task of the university. This is not the place, however, for going into the responsibility of the university more deeply, it would call up a whole complexity of problems that fall outside the framework of our theme, and which we therefore need not discuss.

For the possible remissness of the university does not release the student of science from his own sense of responsibility, a responsibility which very understandably rests heavily on many now that a large part of fundamental scientific research seems wholly aimed at discovering new military means of destruction, research-work which over and above this is largely carried out in secrecy which would seem gravely to impair the freedom of science. The disturbing question then arises whether one can be justified in co-operating in such a scientific programme.

In the following we shall enter into a few points that seem of fundamental importance for answering this problem. In the first place we feel we should point out that in speaking of atom bombs and simular matters, science is often unjustly accused because people wish to stretch its responsibility beyond its proper limits. Above we spoke about the function of science in technics. We saw that technical science leads to a "humanization" of matter, it is drawn into the human sphere. The picture which this technical world will display therefore reflects, on an e-normously enlarged scale, the picture of man, including his vices and his virtues. His virtues; without technical means no amount of heroism would have been able to save thousands of lives in the bad floods of 1953. His vices: without technical means a modern war would not have the destructive power it to-day possesses.

The problem facing us is not therefore banning the production of atom bombs, etc., but how to banish war. When people wish to destroy other people they will always use and find the most efficient means of destruction. As long as war therefore is part of the human doings science and technics will remain bound up with it. Responsibility for there being war does not rest with the students of science as such, they are only responsible in so far as all people are responsible for it. Because war is after all the institutionalized result of the evil and vice in mankind, not of one person or one group of people but of many in conjunction. Almost no one wishes for war, but who can withstand all those small and big acts of egoism which added up result in war? The tragic thing in all this is that the well-meaning are often forced by the use or threat of force to defend themselves against the evilly disposed. It is needless to say of course that science as such cannot do away with evil and vice in man, but it can curb it. Each branch of science can contribute in its own field towards the banning of war by tracing and perhaps removing the causes which bring about that the factually existing evil in man ends in war. Law and economics, physical science and technical science, psychology and sociology, philosophy and theology, they can all furnish their specific contribution towards helping to diminish the tension a-mong the nations. For instance it is not unthinkable that the progress made by physical science and the possibility following therefrom of making ever deadlier and more drastic means of destruction, will exercise a wholesome influence on the will to peace among the nations.

But let us return to the sense of responsibility resting on the student of science who co-operates in developing those means of destruction and who can perhaps hope they will banish war, but can he be certain of that? How can he know they will never be used?

In defining the responsibility of the student of science, two principles should be laid down. The first is that the student of science is not responsible for the abuse to which his knowledge may be put. With every fresh piece of knowledge the possibility is given to use it for good or evil. To guard against abuse is not the direct responsibility of the student of science, however. Even if he foresees possible abuse then that yet need not prevent him from continuing his scientific research work. If it were to do so then every outward human activity would become a thing that was not permitted because abuse is always possible.

The second principle that should guide us is that nobody, and this of course also holds good for the student of science, would give his direct co-operation for evil. If he does so nevertheless, then he is partly responsible for this evil. This principle will naturally hardly be of any importance as long as we have to do with fundamental scientific research, for instance research of nuclear energy, but it is of great importance when the research is directed directly at the production of means of destruction. If one has practical certainty, that these means will not be used to prevent injustice, but to inflict it, one may not render one's assistance. Such instances occurred among German researchers in the last world war.

If one has practical certainty on the other hand that the destructive means will be used solely and only to prevent injustice, then co-operation is not only permitted it may even under certain circumstances be one's duty. It should be noted, however, that there may be situations in which the means are worse than the complaint, in other words destructive means may be discovered that are so terrible that one should rather suffer injustice than have recourse to the means in question. Even in this case, however, it may be necessary to work for such contrivances if only to be able to study possible means of defense. Over and above this it seems probable that if more than one party is in possession of such arms the chance that they will be used is less great than if they are in the hands of only one party. But all this leads us away from the responsibility of exact science.

If now, at the end of our survey, we look over the whole once more, then the conclusion must be that the primary and most important responsibility resting on the student of science must be looked for in his striving after truth, in his determining with the means of his special branch of science, what really is. This after all is his specific responsibility. Every other responsibility resting on him he has in common with other people. I should like to add, however, that the high worth rightly belonging to science, must leave its mark on the whole of the activities of the student of science. It rightly arouses a greater sense of aversion when a scientifically trained man abuses his knowledge than when just anybody does so. One may accept false information and advertisement if need be from a political party or a manufacturer, but one does not do so if this happens on the part of scientifically trained people. The high mission entrusted to man by God: the intellectual fathoming of creation thus making it more instrumental to man, finds its point of culmination in the pursuit of science. This high mission suffers no falsification. Corruptio optima pessima.

SUMMARY OF THE DEBATE ABOUT PROF. A. G. M. VAN MELSEN'S PRE-ADVICE ABOUT "THE RESPONSIBILITY OF THE SCIENTIST"

by D. WIERSMA

Prof. Wiersma brings forward as his first point that the concept freedom is insufficiently defined in the pre-advice. It means in the connection under consideration the absence of compulsion from without.

Also as regards the conception of responsibility he believes that the pre-adviser takes things in too wide a sense, for what it all turns on is moral responsibility only.

Prof. Wiersma concurs with the distinction of the three different fields of responsibility of the student of science, as well as with what is said about it as regards the field of the pursuit of pure science and that of the diffusion of science. The difficulties regarding the responsibilities in the application of science are the greatest. It is the debater's opinion that here one should put oneself on a standpoint that is purely a matter of principle, and that one should not be guided by opportunist motives. If the application of the military means that have been discovered through scientific research are immoral – and this is always so, since we shall always strike the innocent and destroy cultural values for the sake of the subjective advantage of one group, however large it may be – then all researchers, to whatever group or party they may belong, should refrain from putting those means at the public disposal.

SUMMARY OF PROF. VAN MELSEN'S ANSWER
TO PROF. WIERSMA

The pre-adviser naturally has no objection to Prof. Wiersma's definition of freedom except in so far as it is a negative one. The philosopher has to try and penetrate to the positive conceptual content, which has consequently been attempted in the pre-advice. He does not however, wish to go further into this problem at the moment, but he wishes to confine himself to the most important point of difference which came to the fore in the debater's argumentation namely his fundamental rejection of cooperation in any research connected with the development of military means of destruction. The reason that the pre-adviser has not been convinced by the debater's argument lies in the debater's premise that in a war it would turn on subjective advantage of one group.

If this premise were correct, he would accept the debater's consequence, but he doubts its correctness.

To his idea the Dutch people in their struggle against the occupier did not only stand for their subjective advantage but above all they wanted to resist injustice and the impairment of what was sacred to them.

It is this experience that touched every one of us so deeply, that makes it impossible for the pre-adviser to subscribe it the debater's premise and consequently also to his conclusion.

SUMMARY OF THE DEBATE ABOUT PROF. A. G. M. VAN MELSEN'S PRE-ADVICE ABOUT "THE RESPONSIBILITY OF THE SCIENTIST"

by C. J. DIPPEL

Dr Dippel starts by stating that he agrees with the pre-advice in his first approximation to it. He also believes that in many cases it is sociologically demonstrable that the presence of incorruptible scientists has a benificial effect on society.

His objection against the pre-advice is therefore not so much directed against what is said in it, but against what is *not* said in it.

He particularly misses a discussion of the responsibility the scientist bears for integrating his knowledge into the complexity of life, and for anticipating the social results of his scientific finds.

These are perhaps problems that are unsolvable at the moment, the responsibility of the student of science requires however, that he should painfully undergo this unsolvability so that a relationship should arise to the university, theology and society. This raises questions and it is symptomatic that they were eliminated at this congress.

In passing the debater criticizes that part of the pre-advice that treats only of avoiding cooperating with evil directly, since this would diminish people's powers of discrimination for the evil that slinks into society through an endless series of bonds and relationships.

SUMMARY OF PROF. VAN MELSEN'S ANSWER
TO DR DIPPEL

The pre-adviser is grateful for the view of someone who knows all about the practice of applied science and he entirely understands Dr Dippel's objections against the pre-advice. Nevertheless he believes that in the pre-advice especially the responsibility had to be stressed, which belongs to the student of science in that capacity, although at the same time explicitly must be stated that his responsibility was thereby not by any means exhausted. Responsibility that goes further than this the scientist shares with his fellow-creatures as has been repeatedly remarked in the pre-advice. This does not make the responsibility less important but it does make it less specific.

This is the reason that it was only discussed in passing in the pre-advice. Whoever considers the distinction made in the sense of responsibility artificial should remember that the responsibility that goes further than the scientific responsibility cannot be entirely observed with the means peculiar to science, which in turn is connected with the abstract character of science.

The pre-adviser would gladly admit, however, that the student of science, more than anyone else, has the responsibility to bring a sense of responsibility home to people (cf. what is said in the pre-advice about the task of the University). That the debater has more especially brought this side of the problem forward is considered by the pre-adviser to be a fortunate addition to his line of thought.

As regards the debater's last objection, the pre-adviser sticks to his opinion that, regarding the general principles that should guide ethical action, one cannot go further than has been done in the pre-advice, to which he should like to add that careful consideration on the strength of certain principles will rather prevent slackness than further it. It seems to him that slackness is more the result of never thinking about principles than of following too broad principles.

CONCLUDING WORDS

by H. WAGENVOORT

This brings us to the end of our congress. You do not expect or require a lengthy wind-up address of me; it is now time to reflect on what has been discussed and – if possible – to put it into practice. But we may not break up this congress without a word of thanks to all who cooperated in making it a success. I can only once again thank the speakers, pre-advisers and debaters; if our discussions were of a high standard then that is in the first place *their* doing. I also wish to thank the members of the Organising Committee for their kind cooperation; they will approve of my mentioning our Secretary collegue Schouten in this connection, who spared himself no pains, and who probably did not realise what it would entail, when, as secretary of the Board of Vice-Chancellors, he took the organisation of this congress upon himself.

We very particularly wish to proffer our thanks to the national student organisation "World University Service", and especially its Delft section. Its members did an enormous amount of work, of the most thankless sort. The most thankless? Perhaps not after all, that is if the Vice-Chancellors of our Universities have had their eyes opened, if that was still necessary, to the importance of this organisation, which among other things aims at going into the problems of university teaching.

I am of course also thinking of the women students, who yesterday in the interval saw to the tea and no less of the stewards, who with so much ease and grace saw to it that things ran smootly. The management of the Zoological Gardens and their officials are also entitled to our gratitude for their loyal and generous cooperation.

And finally, Ladies and Gentlemen, a sincere word of thanks to you all for your active presence. We are now about to face freedom and restriction once more, albeit with a renewed and intensified realisation of our responsibility.

With this wish I declare this congress to be closed.

REMARKS, DR C. J. DIPPEL WOULD HAVE WISHED TO MAKE AS A DEBATER WITH REGARD TO THE PRE-ADVICE BY PROF. W. F. WERTHEIM ON "ASPECTS OF FREEDOM AND RESTRICTION IN SCIENCE", IF THERE HAD BEEN TIME

There is no science which is so in danger of creating a wrong picture through half truths as sociology. For this reason I wish to give some additional data – without thereby repudiating anything Prof. Wertheim wrote – which to my mind may influence the bearing of his theses and the interpretation of his statements.

I should like to do this with regard to his appraisement of patents and manipulations of patents. I could then demonstrate that patents further the freedom of science and that the introduction of science in modern industry has caused a forcible development with the sociological extremely interesting result that the function and value of patents is greatly changing, so that in the U.S. for instance the licence rights are very easily relinquished in a reciprocity system against very low charges (2% licence right is very usual there among the big concerns).

But since Prof. Wertheim will in any case talk in his explanatory comments on the long-life of light-bulbs as an instance of frustration in science that is to the disadvantage of the consumer, I want to confine myself to that subject. There are three groups of interested parties: the private consumer (who asks for long-life plus quality), the supplier of electricity (who originally had an interest in a high consumption of current), the producer of the light-bulbs (who has an interest in large sales and high quality for keeping up the market). One might add as an fourth group the municipal electrical works and gas works which are both consumer and producer in one.

It is quite definitely the development of science in the hands of the light-bulb producers which made it possible to unite all the parties, which originally seemed to have contradictory interests, around jointly accepted regulations as to quality and life, because this science brought forward a certain amount of knowledge before which all the parties have to submit, both on account of the coercive character of this knowledge as well as on account of the fact that this knowledge is generally accessible, so also for the competitor. It has namely already long been proved that if the cost of a qualified light-unit (i e. price of the bulb plus the cost of the electricity for supplying this unit via the bulb) is set against the life of the bulb, then a curve showing a minimum is obtained. This minimum used to be somewhere around 1000 hours. In consultation with government committees and institutions like the Kema in Holland, etc., the testing has since long been based in the various countries on a certain quality which, together with other characteristics, also has a life of about 1000 hours. Germany led the way in 1928, in 1930 it was the Brit. Eng. Standards Association, in 1932 Belgium, 1934 it was Holland with a report of the *Cie. van Bijstand v.h. Gloeilampenvraagstuk* (Advisory Company for the Light-Bulbs question) in which the choice of 1000 hours for the retail consumer was made, and in 1936 it was Russia which came out with explanatory notes to exactly the same effect as had already been given in Western Europe. In any case neither the Phoebus concern nor Dr Philips was a party to these Russian regulations. As soon as such discriptions of quality have been fixed, however, supplying of bulbs that are underrun implies that the individual is being misled and that the community is being defrauded. This is all the more so when one knows that at the moment it would be recommendable, on account of the current having become dearer and the costprice of the bulb lower in many countries (among others Holland), to put the life of the bulb considerably lower than 1000 hours. [1].

Dr Philips has greatly stimulated the introduction of science in industry. He has also based his commercial tactics on science to his own advantage. As early as 1934 he stimulated the development of sodium and mercury lamps, where the question of long-life is quite different and more favourable (more favourable, i.e. makes a considerably longer life possible, even when the above-mentioned electricity costs are taken into account). He often visited the scientific laboratory in his capacity of employer. Perhaps in distinction to sociologists (who cannot accept a certain amount of science and a certain amount of parallelism on this point be-

---

[1] Prof. Wertheim is invited to come and consult about a few points with the scientific staff of the *N.V. Philips' Gloeilampenfabriek* (Philips Light-bulb Factory and Co. Ltd) at Eindhoven, in the company of a physicist to be named by him.

tween the interests of the employers and society because of an old-established myth) scientific researchers can very well understand that he received an honorary doctorate. It is partly through his instrumentality namely that objective science here brought about a frustration of classic commercialist methods. This struggle between science and commercialism is still of everyday occurrence, and is different in every field. The development is sociologically decidedly more interesting, however, than the usual generalizing suggestions about frustration. Undoubtedly it is possible to suspect the existence of powerfully frustrating forces, although in my experience evidence for frustration can only be provided if one proves in practice that for instance the supposedly frustrated plan can be economically maintained. In closer analysis it will often appear that it is freedom which causes frustration, because with the granting of freedom the employer has the right to combine the *smallest* risk with the strongest sense of pleasure among the consumers. With the abolishment of freedom other modes of frustration are called into being under a dictatorship, however. This might be a reason why this idea of frustration is unsuitable for describing the situation entailed by freedom and restriction in science. The intention I have in mind with these additional remarks is to show at a given point that the progressive march of science may have a purifying effect on manipulations threatening the individual and society, and that via science the properly understood interests of the employers may very well appear to coincide with the interests of society. The question what science does and what science could do requires *specifically* expert solutions, not generalizing qualifications or suggestive ideas. The scientific research worker with a sense of reponsibility has here in industry a far from hopeless, but rather a hopeful task.

PROF. WERTHEIM'S WRITTEN ANSWER TO DR DIPPEL'S REMARKS

Whoever reads Dr Dippel's remarks would hardly be likely to surmise that they have not been inspired by some abstract, generalising argument, but by a very concrete example, namely Mr A. F. Philips' letter printed on p. 94 of my pre-advice. In this letter to Mr Clark Minor of the General Electric, dated 1934, the disadvantage of too high voltages are judged *solely* regarding their effect on the total sales. There is question of "strenuous efforts we made to emerge from a period of long-life lamps", and of the serious consequences too high voltages might have for the Phoebus combination. The interests of the consumer are not so much as alluded to. It is in any case important that the authenticily of Philips' letter is not denied by Dr Dippel.

It will be in vain to look for utterances in my pre-advice that throw doubts on the great scientific achievements of Dr Philips, as Dr Dippel tries to suggest. Nor will a denial be found in my pre-advice of Dr Dippel's proposition that in concrete cases via science the interests of the employers *may* (the italics are mine, W) prove to coincide with the interests of society[1]. On the contrary, notwithstanding the great difficulty in getting positive proof, one will find quite a large number of concrete examples of frustration, as well as some very cautious and relativist observations on the phenomenon of frustration in general.

But it is Dr Dippel who generalises in an unwarrantable way, when he lays down that the lamp producer is interested "in large sales and *high* (italics are mine, W) quality for keeping up the market". Because Dr Dippel knows as well as I do that the element of free competition in the lamp industry is only to be found in a very limited measure, and that in the past this was even less so on account of patents and monopoly positions. For this reason Dr Dippel's statement that science "is generally accessible, *so also for the competitor*" (italics are mine, W) is also an incorrect generalisation and so only a half truth.

And finally it is down right incorrect that "*all* (italics are mine, W.) the parties, which originally seemed to have contradictory

---

[1] My statement was: "In practice the interests of the employers by no means always coincide with those of the consumers" (p. 93).

interests" are united around "jointly" accepted regulations as to life and quality. Because Dr Dippel is bound to know that the consumer, who at that time was not yet organised in a consumers union, was not a party to these decisions. A great deal more is therefore necessary than Dr Dippel brings forward to prove that, notwithstanding the apparent semblance to the contrary, the policy accepted by Mr Philips and Mr Minor in 1934, although entirely argued from the point of view of the interests of their industry, was in casu partly fixed on the strength of the interests of the consumer. How little the interests of the consumer counted in the circles of the General Electric appears clearly from an American prosecutor's report of 1944, quoted by William Kapp, "The Social Costs of Private Enterprise", 1950. For in it a proposal made by one of the directors of this concern is mentioned to limit "the life of flashlight lamps from the basis on which one was supposed to outlast three batteries to a point where the life of the lamp and the life of the battery under service conditions would be approximately equal.

If this were done we estimate that it would result in increasing our flashlight business approximately 60 per cent. We can see no logical reason, either from our view point or that of the battery manufacturer, why such a change should not be made at this time".

As far as I am aware I have nowhere in my pre-advice defended any "old established myth", but have confined myself to concrete, published facts. But I have no use either for a new myth about the purifying influence of science on the modern type of employer and about classic commercialist methods becoming impossible through "objective science". The Amsterdam section of the *Verbond van Wetenschappelijke Onderzoekers* (the Association of Scientific Research Workers) has collected too much incontestable material for that (which has partly already found its way into my pre-advice, and will partly as yet be published). From this material it appears most clearly that commercialist methods, whether classic or not, are still as much able as ever to injure considerably the interests of the consumer.

Dr Dippel mistakenly tries to refute the value of my arguments by making it appear as if my argumentation is that of a sociologist who has never seen the inside of scientific laboratories. As I have already shown on p. 93 of my pre-advice, my arguments, as well as the examples, are based in great part on an investigation carried out by the *V.W.O.*, in which representatives of the natural sciences play an important part. Although I bear the full responsibility of the formulation and rendering of those results in my pre-advice, I have yet in that way had the opportunity in numerous talks with researchers of the natural sciences exhaustively to test their train of thought, as well as the concrete material.

These scientific research workers also possess a sense of responsibility and consider their task in industry to be far from hopeless. But they consider it their specific duty to remain critical in regard to their own industrial concern too, and not to allow themselves on account of a strong "psychological tie with the industry" to think the interests of their concern and of society to be identical.

I shall gratefully accept Dr Dippel's invitation if it reaches me on behalf of *N.V. Philips' Gloeilampenfabriek,* and if it contains a satisfactory definition of what is to be the subject of discussion and investigation.